T0074913

Survival Analysis with Python

Survival Analysis with Python

Avishek Nag

CRC Press is an imprint of the
Taylor & Francis Group, an **informa** business
AN AUERBACH BOOK

First Edition published 2022
by CRC Press
6000 Broken Sound Parkway NW, Suite 300, Boca Raton, FL 33487-2742
and by CRC Press

2 Park Square, Milton Park, Abingdon, Oxon, OX14 4RN

CRC Press is an imprint of Taylor & Francis Group, LLC

Library of Congress Cataloging-in-Publication Data
A catalog record for this title has been requested

ISBN: 978-1-032-14826-7 (hbk)
ISBN: 978-1-032-07367-5 (pbk)
ISBN: 978-1-003-25549-9 (ebk)

DOI: 10.1201/9781003255499

Typeset in Adobe Garamond
by KnowledgeWorks Global Ltd.

Contents

Preface

As a subject, survival analysis has a long history. Most recent interests in this area are observed primarily in medical science (biostatistics, mainly) where effects of different factors on human life expectancy after a disease are analyzed. Other areas are industrial-life testing, analysis of extreme natural calamities, and some topics in physics, social science, etc. It is a statistical approach to study any system's (biological, physical, mechanical, social, etc.) expected time until it works reliably. To understand it thoroughly, good knowledge of probability, statistics, and calculus is a prerequisite.

The motivation for writing this book came from the need of resources combining the theory as well as very practical approaches in a single place. Nowadays, Python has almost been a de facto standard for any analytical, data science, or statistical computation-related topics. Moreover, as of now, there is almost no book on survival analysis in combination with *Python*. Being a subject that is very mathematical in nature, I tried to make this book work as a bridge between the complex theories and their implementations with code. For instance, theories like mathematical derivations of Gumbel density & survival function and very practical topics like choosing covariates in an iterative manner are both covered.

In my opinion, this book should guide software engineers, data science practitioners, and statisticians to get a flavor of the subject from both theoretical and practical points of view. It may also help students for a short half-semester course on related topics.

Note: Readers are requested to install Python 3.x with *lifelines* library. I recommend using Jupyter notebook for ease of use.

About the Author

Avishek Nag has a Master's in Data Analytics & Machine Learning from Birla Institute of Technology (Pilani, India) and a Bachelor's in Computer Science. He has work experience in different renowned companies, such as VMware, Cognizant, Cisco, Mobile Iron, etc. He started his career as a software engineer and later moved to Machine Learning (ML)/Data Science. He has been an analytics practitioner for several years now, specializing in statistical methods, machine learning, and natural language processing. He has experience in designing end-to-end Machine Learning systems, driving Data Science/ML initiatives starting from inception to production in multiple organizations. He also acts as a key contributor for several online publications like Towards AI, Towards Data Science, etc., which are focusing on Artificial Intelligence and Data Science. His blogs can be found at https://medium.com/@avisheknag17. He is passionate about authoring books & doing research in several areas of applied Math/Stat/Probability/Computational Science. His interest areas are Mathematical Optimization, Survival Analysis, Computational Finance, Chaotic Dynamical Systems, etc.

Chapter 1

Introduction

We will start our discussion with a few events that can be observed: death of a person due to a disease, attrition of an employee from an organization and incident of a natural calamity (earthquake or flood). All these examples are from completely different domains, but they have a common thing: time or, better to say, time until an event occurs. Time is crucial in all these situations. If we know beforehand that a certain event may occur at any specific time, then a lot of lives and resources can be saved. *Survival analysis* is defined as a collection of statistical longitudinal data analysis techniques where time is a major factor. It is utilized in biology, medicine, engineering, marketing, social sciences or behavioral sciences. Survival analysis is also sometimes named as *reliability theory* under *operations research* or *engineering*. It is a complex subject and the reader would need expertise in probability, statistics, calculus and optimization to grasp it fully.

In this chapter, we will explore some basic concepts of survival analysis, nomenclatures and sample datasets.

Concept of Failure Time

We have already talked about event. In general, survival analysis deals with the events related to failure. And failure off course can occur one or more time for any subject. For the topics discussed in this book it is assumed that failure occurs only once for a subject. We will be using the term *subject* throughout this book to represent the entity which is going through some phases and the failure (or the event) is attached to it. A ***subject*** may be a person, a machine, a river, and even an entire geographic region. There are numerous use cases where survival analysis can be applied to find out chances of event occurrence. Some of them are:

DOI: 10.1201/9781003255499-1

- Death of a person by any disease
- Suicide
- Failure of machine tools
- Attrition of employees from organization
- Divorce
- Occurrence any natural catastrophe (flood, earthquake, volcanic eruption, etc.)

In this book, we will be discussing mostly about the death by disease use cases, as survival analysis finds its usage in these cases mostly. Death by disease use case is mostly analyzed in case of *drug* development, where survival analysis plays a crucial role to identify the right drug by comparative study of several options.

We are talking about *time* a lot. But what does it signify? By *time*, we mean years, months, weeks or days from the beginning of analysis of the data until an ***event*** (like death, exit of an employee, earthquake, etc.) occurs. As said earlier, *event* is also termed as ***failure***. So, time taken till failure is referred to as the ***failure time*** or ***survival time***. Time may not be a physical unit always; there are cases where it can be used as a logical indicator. Below points are needed to be taken care of before defining a time scale:

- Origin of the time must be unambiguously defined.
- The scale for measuring the time difference must be defined.
- Definition of failure must be clear.

Concept of Survival

When we speak about ***survival***, we mean probabilities. Probability of not occurring an event till some time can be taken as survival probability. In other words, probability of an event occurrence after a certain time is survival probability. For example, when we say survival probability of a heart patient at age 71 is 0.23, it means that the patient will survive at least till age 71 and there is a probability 0.23 that he/she will keep surviving after 71. Age is a time scale here. Similarly, there could be a probability 0.40 that he/she will survive after 50. Reason is clear. At younger age, chances of collapsing by a heart attack is less and thus survival probability will be higher. So, we can have a survival probability distribution over random variable *time* (here *age*) like below:

Table 1.1 A Sample Survival Probability Distribution

Time (Age)	40	45	50	60	65	70
Survival Probability	0.51	0.42	0.38	0.36	0.28	0.24

One of the purposes of survival analysis is to find out this probability distribution. A lot of other domain-specific statistical inferences can also be drawn from this. It can be observed that survival probability decreases over time. It is a very important feature of distribution. We will discuss it in greater detail in Chapter 2. Like heart patient use case, the same analysis can be done for employee attrition of an organization. The purpose is to find out survival probability distribution of employee exit at various times after he/she joins there. Interesting part is that the term *survival* is very generic here. It should not necessarily always mean *saving yourself from something.* It is not also always related to disease, patients or healthcare. *Survival* means non-occurrence of an event till some time. Events could either be any one from the list as discussed in the section 'Concept of Failure Time' or something else.

Censoring

Most survival analyses must consider a very important analytical problem called ***censoring.*** It is caused by not observing some subjects for the full time till failure (or event). We will consider the patient use case to understand the problem easily. We have records for patients dying from heart attack, but in some situations, it may not be possible to mark the exact time of death. Patients might have died either before or after the marked time value. In this case, it is said that the data is ***censored.*** In medical diagnosis, a ***study*** involves regular follow-ups with the doctor by the patient. The doctor starts taking notes about the patient's health condition in each follow-up and schedules the next date. Problem occurs when the patient dies in between, after the end of or before the study, and hence ***censoring*** occurs. There can be three primary reasons for this:

1. Patient does not have the event (death) before the study ends.
2. Patient left follow-up during the study period.
3. Patient died in the study period.

In all above cases, true survival time is not equal to the observed survival time, as the actual time could not be marked. Depending on three situations, there can be corresponding three types of censoring as discussed next.

Right Censoring

Right censoring happens when study ends but no event is observed. The patient did regular follow-ups in the study period. The event might have occurred after the study, but actual time could not be noted and hence it is censored. The below diagram explains the situation.

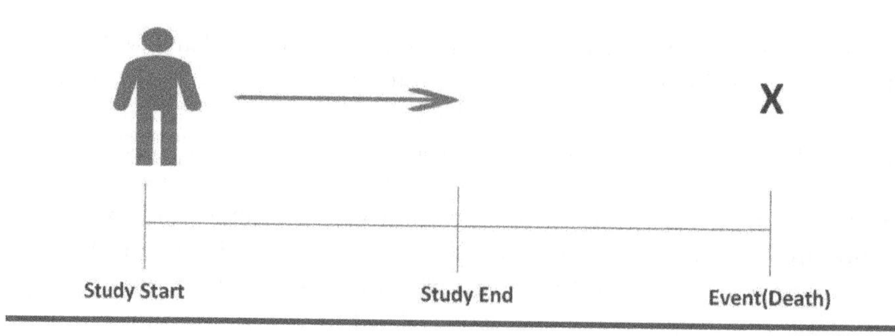

Figure 1.1 **Event sequence for right censoring.**

For this situation, complete survival time interval has been cut-off at the right side of the observed time (study end). Hence, true survival time is greater or equal to the observed time. It is the most popular censoring mechanism. For all the models in this book, we will assume ***right censoring*** and design it accordingly.

Left Censoring

Left censoring happens when event has already occurred before the start of the study. The following diagram depicts the situation.

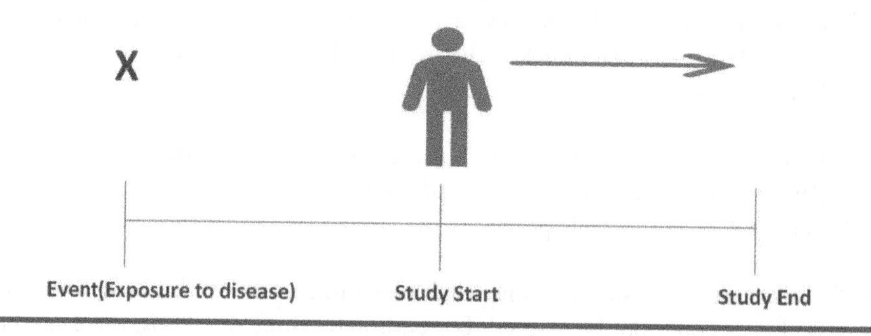

Figure 1.2 **Event sequence for left censoring.**

For this case, true survival time is less than or equal to the observed time (study end). It is useful in modeling the use cases where patients are doing regular follow-ups and the exposure of the disease is considered as an event. For example, if people are being tested for HIV positive over a period, then the first exposure (i.e., first test which comes as positive) is noted as failure (or event). It may happen that for some people, exposure happened before the study (i.e., they are already HIV positive), but exact time of it could not be noted as the study did not start then.

Interval Censoring

Interval censoring happens when event occurs within the study period in between two possible time limits, and as usual actual time could not be noted. The below diagram explains the situation.

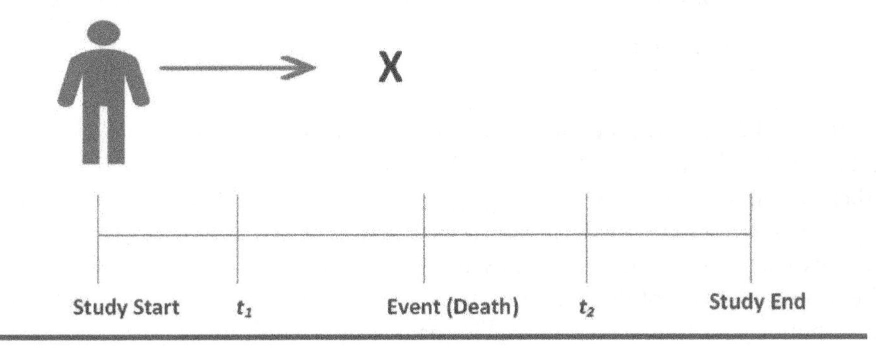

Figure 1.3 **Event sequence for interval censoring.**

We can see that event occurred somewhere between time t_1 and t_2. This scheme is applicable to both death and exposure to disease events. For example, a person may be detected as HIV negative at time t_1, but within the study period, he/she might have got exposure and testing at time t_2 identified him/her as HIV positive. Interval censoring is a combination of both left and right censoring with one limit as infinite.

Sample Dataset Structure

Any survival analysis-related dataset follows a typical structure. From our discussion so far, we can easily understand that time information should be there at first. Second, event status should be there. It is also known as ***censoring status***. It should be either a Boolean or a status indicator (0 or 1). Value 1 indicates occurrence of the event and it is not censored, whereas a 0 value says non-occurrence and it is censored. Survival analysis dataset often contains information about the subject under study. For heart attack use case as discussed in earlier sections, dataset may have attributes of the patient. So, a typical structure may look like below:

Table 1.2 A Hypothetical Dataset for Survival Analysis

Person ID	*Group*	*Time*	*Status*	*A*	*B*	*C*	*D*
P_1	Control	T_1	0	A_1	B_1	C_1	D_1
P_2	Treatment	T_2	0	A_2	B_2	C_2	D_2
P_1	Control	T_3	1	A_1	B_1	C_1	D_1
P_3	Treatment	T_4	1	A_3	B_3	C_3	D_3
P_4	Control	T_4	0	A_4	B_4	C_4	D_4
P_2	Treatment	T_5	1	A_2	B_2	C_2	D_2

A, *B*, *C* and *D* are attributes of the patient (could be blood pressure, weight, or any other medical/health-related statistics). These records form a distribution over time. There are multiple records for a single person for few cases. For example, person P_1 has two entries at time T_1 and T_3, respectively. As it is a distribution over time, records may contain same subject's status information over different time values. Remember that one of the purposes of survival analysis is to find out a probability distribution over time for a subject. For that reason, records may also contain (not necessarily it should) same subjects' information over different time values. It is quite analogous to a time series database. One of the practical examples of this type of dataset is given below:

	Unnamed: 0	trt	celltype	time	status	karno	diagtime	age	prior
0	1	1	squamous	72	1	60	7	69	0
1	2	1	squamous	411	1	70	5	64	10
2	3	1	squamous	228	1	60	3	38	0
3	4	1	squamous	126	1	60	9	63	10
4	5	1	squamous	118	1	70	11	65	10

Figure 1.4 A practical dataset (discussed later) for survival analysis.

We will discuss about the source and meaning of the dataset in later chapters of the book. Next, we will discuss a few terminologies.

Control and Treatment Group

There is a *group* column in the sample dataset structure. Patients are divided into two groups: ***control*** and ***treatment***. There may be any other name for these groups (for example, Group 1 and 2). But specifically, *control* and *treatment* has some purpose in drug development and medicine. We will be using these terms in later chapters also whenever we have groups in the dataset. Generally, survival analysis is done for different groups parallelly to do a comparative study. Consider an example of a new drug for heart disease to be tested at a hospital. Patients are divided into two groups as usual: ***control*** and ***treatment***. Treatment group patients are given the real drug, whereas Control group patients receive ***placebo***, a harmless pill not containing any active ingredients. In this way, patients do not know which patients are receiving the new drug. The purpose of doing this is to eliminate the psychological effects that might impact the result and analysis. Considering heart attack or some symptoms as event, survival analysis is done for two groups. Two probability distributions are obtained as result and they are compared with each other. If the drug is effective, then there will be less chances of events and thus probability distribution of ***treatment*** group will be significantly different from the ***control*** one. As a result, patients who are given the drug will have better survival probabilities. We will discuss more about comparing two distributions in Chapter 4. Instead of one control and one treatment group, we can also do survival analysis on two treatment groups. Its purpose will be to compare effects of two different drugs. If survival distributions are statistically similar, then two drugs are declared similar and any one can be chosen.

Risk Set

There are six records in the sample dataset structure discussed above. Observed that there are information about four persons (P_1, P_2, P_3 and P_4) only. Some records are repetition of same persons with a different time and status values. So basically, we have a unique set of four records. This set is known as ***risk set***. We will be using this term in subsequent chapters.

Comparison with Regression

Problem statements of survival analysis sound quite similar to regression problems. A regression model can be built considering *'Time to failure'* (i.e., time elapsed from present to the actual event) as target variable. That can still at least solve the problem by giving the predicted time of the next event. If that is the case, then why would we need special technique like survival analysis? Main reason is *censoring*. We need to know the exact time values of event to fit a standard regression model. Censoring prevents us there. It may give poor result because of that.

Another reason is the distribution of time to event. If the *'Time to failure'* is normally distributed, we still could use a regression (or logistic regression model). But that is not the case. Distribution of time is either unknown or not normal at all. We will discuss in detail in Chapter 2.

Chapter 2

General Theory of Survival Analysis

In Chapter 1, we discussed about different nomenclatures of survival analysis. We understood that it is an analysis of events which keeps occurring over a period. We can express these events as probability distribution over time. Here, time T is the random variable. Our objective is to find the probability of the specified event for a subject after specified time t. It gives us the idea of chances of future events. To analyze it further, we need a few metrics as discussed next.

Survival Function

Survival function returns the probability of an event occurring after time t. Mathematically,

$$S(t) = P(T > t),\ 0 < t < \infty$$

Alternatively, it can be said that $S(t)$ gives us the probability of a subject surviving after time t. $S(t)$ is nothing but a probability distribution over time. Theoretically, t ranges from 0 to infinity, and of course, $S(t)$ will have values from 0 to 1. Ideally, *survival function* is represented by a decreasing smooth curve which begins at $S(t) = 1$ at $t = 0$.

It looks like below:

DOI: 10.1201/9781003255499-2

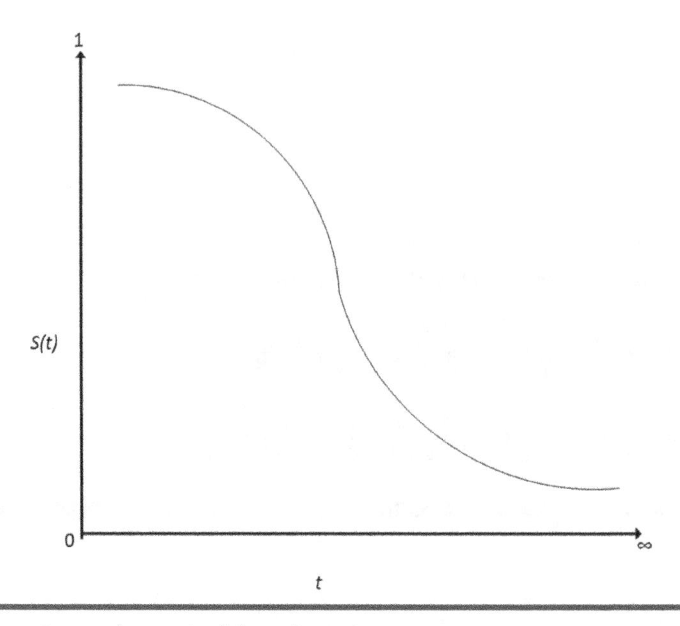

Figure 2.1 **A sample survival function curve.**

The above curve is mostly true for pure theoretical models. In practice, curves generated from real datasets look more broken and stepwise. One example is shown below:

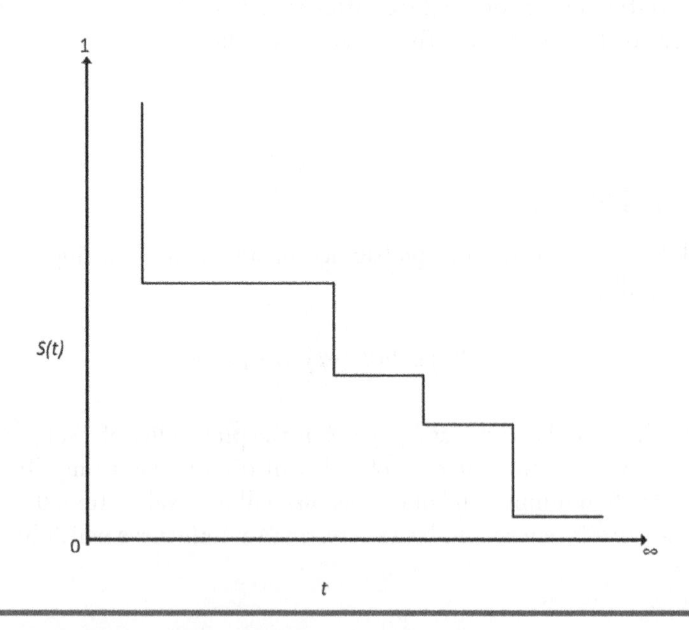

Figure 2.2 **A sample non-parametric survival curve.**

One important property of survival function is that it is monotonically decreasing, i.e.,

$$S(t_1)\langle S(t_2) \text{ where } t_1\rangle\, t_2$$

On a practical note, survival chances from an event tend to decrease over time. For example, chances of surviving of a cancer patient (from event 'Death') decreases as the time passes.

In survival analysis, we could have simply computed the chance of an event from the historical data, but we don't do that. Instead, we compute things in an indirect manner, i.e., computing chances of survival. Basically, probability of an event at certain time is equivalent to the probability of surviving after that time. They both mean the same thing. But we never compute probability of an event at any time. *Censoring* is the reason for that. As discussed in Chapter 1, due to censoring, it is quite impossible to know the exact time of an event. If the accurate time itself is not known, for obvious reasons, probability of an event cannot be computed directly. Hence, the other path is chosen, i.e., probability of surviving from an event after certain time which is the complement of probability of an event. A practical example will explain it better. Suppose a patient was suffering from cancer and was under regular medical checkup. Suddenly, from one day, he stopped coming to the doctor. Doctor came to know later that the patient died in between but the exact date remained unknown. If t_1 is the last time he visited the doctor and t_2 is the time the news of his death came to the doctor, then the actual time of death is somewhere between t_1 & t_2 and is unknown. If a model is to be built from this information, then it will consider computing probabilities of survival after t_1 or t_2 rather than computing the direct event probabilities at those. You may think, why are we still making it complex by probabilistic approach? Why can't we just predict the time of the event directly? Uncertainty is the answer, and it is caused primarily by censoring. As we never know the actual time of event, we go by chances and *Theory of probability* is the best option we are left with.

But there is one argument that may work against Survival Analysis. Consider the same patient-doctor example. Let t_{Death} be the actual time when the patient died, and it is still unknown to the doctor. Magnitude of the difference ($t_2 - t_{Death}$) matters a lot here. And of course, the difference is caused by the uncertainty about t_{Death}. If the overall system is too much sensitive about this difference, then doing Survival Analysis makes sense otherwise it is an overkill. Secondly, if ($t_2 - t_{Death}$) is too small then we can ignore it and Survival Analysis is not needed there. Alternatively, a regression model can be built with *'Time to failure'* (till time t_2) as target variable. Then it can predict after how much time the next event will occur. Use cases like human-life analysis, drug testing, accurate prediction of failure time of some complex machines where huge cost of maintenance is involved are perfect candidates for doing Survival Analysis.

From the above discussions, three pre-conditions of performing Survival Analysis can be concluded as described below:

a. Chances of Survival always decrease over time
b. Exact time of events are unknown or censored
c. High sensitivity about the uncertainty of actual event timings.

Hazard Function

Conceptually, *hazard* is opposite to *survival*. It is the event rate or death rate (in case of analysis of impact of diseases on humans). It can also be termed as *failure rate*. A hazard function is represented as $h(t)$.

Mathematically,

$$h_T(t) = \lim \frac{P(t < T < t + \delta \mid T < t)}{\delta}$$

$$\delta \to 0 \tag{2.1}$$

So, hazard at time t is potential per unit time for the event to occur given that the subject has survived till t. Basically, it the rate of event at time t. It is clear from the expression and definition that hazard is a rate rather than being a probability. Its value ranges from zero to infinity.

The following diagram explains the relationship between hazard and survival well.

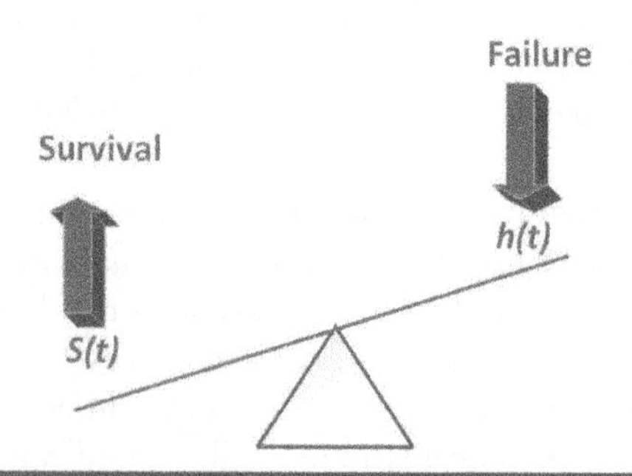

Figure 2.3 **Survival vs hazard function comparison.**

Hazard and survival functions are just two ways of specifying a survival distribution.

Analysis of Relationships

Survival function can be written like below:

$$S(t) = P(T > t) = 1 - P(T \leq t)$$

Survival probability of a subject at time t gives the probability of that surviving after t.

Now, by the law of the probability, $P(T <= t)$ gives nothing but the distribution function of a probability distribution. By theory, it is $P(-\infty < T <= t)$. But in our case, as time scale t starts with zero it is equivalent to $P(0 < T <= t)$.

$P(T <= t)$ is also known as *cumulative risk function* and denoted by $F(t)$. So, we can see that

$$F(t) = 1 - S(t)$$

From theory, we know that probability density function (denoted by $f(t)$) is the derivative of the distribution function. So,

$$f(t) = \frac{d}{dt} F(t)$$

$$= \frac{d}{dt} [1 - S(t)]$$

$$= -\frac{d}{dt} S(t) \tag{2.2}$$

Density function is the negative of the derivative of survival function.

From Equation (2.1), by the theorem of conditional probability and omitting suffix T, we get expression of hazard function as

$$h(t) = \frac{f(t)}{S(t)} \tag{2.3}$$

Basically, the hazard at time t is the probability that an event occurs in the very short neighborhood of time t divided by the probability that the subject is alive at time t. Being a ratio of two probabilities, *hazard* it is a rate rather than a probability itself.

Cumulative hazard function is the total hazard from zero to time t. It is given by

$$H(t) = \int_0^t h(u)\,du$$

Now, we will see how hazard and survival function can be related without using density function.

We know, that

$$\frac{d}{dx}\log(x) = \frac{1}{x}$$

and

$$\frac{d}{dx}\log\left[f(x)\right] = \frac{1}{f(x)}\frac{d}{dx}f(x)$$

We will use the above derivatives of Equations (2.2) and (2.3) to deduce the following:

$$h(t) = \frac{f(t)}{S(t)}$$

$$= -\frac{1}{S(t)}\frac{d}{dt}S(t)$$

$$= -\frac{d}{dt}\log\left[S(t)\right]$$

So, by integrating both sides, we get

$$\log\left[S(t)\right] = -\int_0^t h(t)dt = -H(t)$$

Hence,

$$S(t) = e^{-H(t)} \tag{2.4}$$

Equation (2.4) is the formal relationship between *cumulative hazard* and *survival* function.

With all above analysis, we can see that survival and hazard are functions of time t. It seems like survival probability of a subject is dependent only on time. For now, assume that this is true. We call it *univariate model*. We will see in Chapter 5 that how other factors can influence survival and hazard function along with time.

Estimating Survival Distribution

We have discussed about survival and hazard functions. Survival distribution is nothing but a probability distribution. Now, the question arises: how to compute or estimate that probability distribution? From statistical theory, it can be modeled using either parametric or non-parametric approach.

Parametric approach goes with a strong assumption and uses pre-defined standard probability distribution. Note that only density function will be assumed over here. Later, survival function or hazard can be derived from density function. Standard probability distributions such as *exponential*, *Weibull*, *Gumbel*, etc. are good candidates for parametric density functions.

For non-parametric approach, none of the pre-defined distributions are used. Rather, a custom estimator is built from the dataset. One of the examples is *Kaplan-Meier* estimator. We will discuss parametric and non-parametric approaches in Chapters 3 and 4, respectively.

Now, let us take an example and build a survival function from a real dataset. We will use *veteran* dataset distributed by R survival package (http://www-eio.upc.edu/~pau/cms/rdata/doc/survival/veteran.html, http://www-eio.upc.edu/~pau/cms/rdata/datasets.html). We will read the dataset first.

```
import pandas as pd

veteran_df = pd.read_csv('data/veteran.csv')
veteran_df.head()
```

	Unnamed: 0	trt	celltype	time	status	karno	diagtime	age	prior
0	1	1	squamous	72	1	60	7	69	0
1	2	1	squamous	411	1	70	5	64	10
2	3	1	squamous	228	1	60	3	38	0
3	4	1	squamous	126	1	60	9	63	10
4	5	1	squamous	118	1	70	11	65	10

Listing 2.1 **Read veteran dataset.**

It is a typical survival analysis dataset and like the structure explained in Chapter 1. It is a dataset containing lung cancer patients' information and their survival status for two different types of medical treatment. It contains *status* and *time* column which are our matter of interest. Column *status* indicates censoring status (patient has survived or not). There are two treatment groups – 1 and 2. We will be interested only in *trt* = 1.

Now, we will use exponential function to model survival distribution (*lifelines* package has *ExponentialFitter*). It is a parametric approach.

```
from lifelines import ExponentialFitter

veteran_df_1 = veteran_df[veteran_df.trt==1]

epf = ExponentialFitter().fit(veteran_df_1['time'], veteran_df_1['status'])
```

Listing 2.2 **Fit exponential model on veteran dataset for treatment number 1.**

fit function takes two arguments as *time* and *status* variables.

Let's now plot the survival and hazard functions.

```
import matplotlib.pyplot  as plt

epf.plot_survival_function(label='survival function').legend()
epf.plot_hazard(label='hazard').legend()
```

Listing 2.3 **Plot survival and hazard functions for the fitted exponential model.**

It produces following:

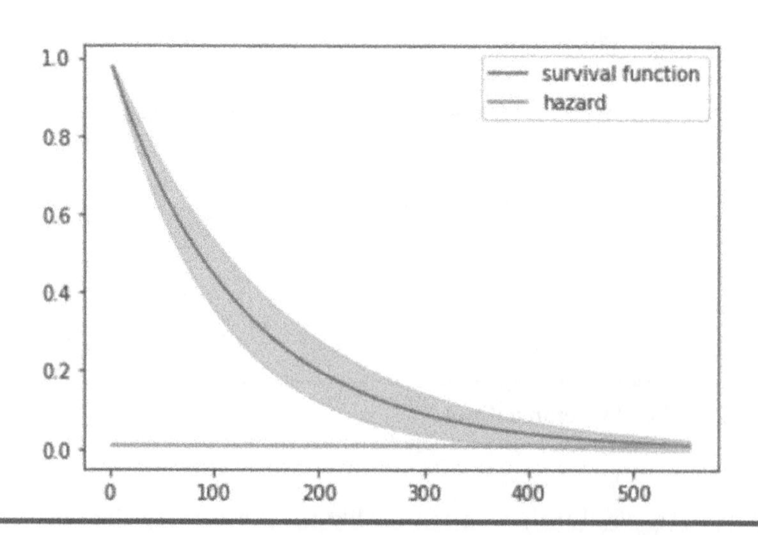

Figure 2.4 **Survival and hazard function plot for veteran dataset.**

We can see that hazard function is constant. Reason will be explained next.

Probability density function (PDF) of *exponential* distribution is given by,

$$f(t) = \rho e^{-\rho t} \text{ (}\rho \text{ is a constant)}$$

From there, we get survival function

$$S(t) = -\int f(t)dt$$
$$= e^{-\rho t} \quad (2.5)$$

And hazard,

$$h(t) = \frac{f(t)}{S(t)} = \rho$$

So, it is proved that hazard function is a constant for *exponential* PDF. It can be interpreted by this statement '*death rate is constant for cancer patients*'.

Predicting Survival Probability

As mentioned earlier, our discussion here will be limited to univariate models; hence, only time t is considered as independent variable. So, to get survival or event probability for any future event at any time instance t, only input to the survival distribution will be the *time* itself. We will see how to get probability of any future event with the test dataset.

We will split the dataset into *train* and *test*.

```
from sklearn.model_selection import train_test_split

train, test = train_test_split(veteran_df_1)
```

Listing 2.4 Split veteran dataset into train and test.

Now build the model using *train* dataset.

```
trained_epf = ExponentialFitter().fit(train['time'], train['status'])
```

Listing 2.5 Fit exponential model on *train* dataset.

Then evaluate it with *test* dataset.

```
estimated_sp = trained_epf.predict(test['time'])
```

Listing 2.6 Predict survival probabilities with the model for test dataset.

estimated_sp contains all estimated survival probabilities. Let us see its contents,

```
51      0.641854
162     0.244527
118     0.358476
278     0.089195
31      0.763750
384     0.035490
151     0.269067
22      0.825911
411     0.028065
10      0.916732
92      0.449397
117     0.361606
8       0.932812
228     0.137761
105     0.401370
82      0.490216
12      0.900930
392     0.033106
```

Figure 2.5 **Estimated survival probabilities.**

It has probabilities of different time instances. For example, at time instance 384, probability of survival is almost 0.035.

Computing Accuracy

How do we know that our estimation of survival probabilities is accurate? There is a technique to compute closeness of probability values of survival function.

From theory, *Brier Score* is a metric which can give accuracy of any probability measure compared to the actual. Mathematically, it is defined as

$$Brier\ Score = \frac{1}{n}\sum_{i=1}^{n}(f_i - o_i)^2$$

where f_i = estimated probability, o_i = actual output (0 or 1)

Actual output will be status indicator (0 or 1). It is analogous to the survival analysis censoring status. Status values as 0 and 1 indicate probability as 0 and 1, respectively. Brier Score is like *mean squared error* of probabilities. A lower value indicates better result.

We will use a custom function to compute Brier Score.

```
import math

def brier_score(actual, estimated):
    n = len(actual)
    error = 0.0
    for i in range(n):
        error = error + math.pow(estimated.iloc[i]-actual.iloc[i],2)
    return error/n
```

Listing 2.7 **Function to compute Brier Score for any survival model.**

Now, to get the accuracy of the exponential model, we need to compare the estimated probabilities and the existing censoring status.

```
brier_score(test['status'], 1.0 - estimated_sp)

0.33087138635556457
```

Listing 2.8 **Use *brier_score* function to compute accuracy of the model.**

Observe that, '*status*' and *estimated_sp* is not directly comparable as the later one is used for survival probability whereas '*status*' gives the event chances for time less than a fixed value. That's why complement of *estimated_sp* (which is 1- *estimated_sp*) has been used.

So, Brier Score of exponential model is around 0.33. This can be improved with a better model. We will see those topics in later chapters.

Mean and Median Survival Time

Mean survival time of a subject is the area under the survival curve from zero to maximum time. It is given by

$$\mu = \int_{0}^{t\max} S(t)\,dt$$

But this will only hold true if the area under the survival curve is finite. It the average survival time for any subject under consideration.

Median survival time at which survival probability is 0.5. But if, the survival distribution is not continuous, then it is taken as the smallest time t for which $S(t)$ <= 0.5 (as there may be multiple values of t for which condition $S(t)$ <= 0.5 may be true).

Chapter 3

Parametric Models

In Chapter 2, we discussed the basic statistical theory of survival analysis. We know that survival and hazard functions are the basic building blocks of analysis. It is super important to estimate these functions properly. We discussed at a high level about parametric and non-parametric approaches. Parametric models are often easier to work with. In traditional machine learning, linear regression, logistic regression and Poisson regression are some of the common parametric models that are used for health sciences domain. These models are nothing but applications of standard statistical distributions which are well-defined and suitable for specific domains. For survival analysis also, we have a series of parametric models such as *exponential, Weibull, Gumbel,* etc. We already saw an example of model fitting with exponential distribution in Chapter 2. In this chapter, we will explore about *Weibull* and *Gumbel* and discuss *exponential* in greater detail.

Maximum Likelihood Estimation (MLE) of Parameters

Each of the probability density functions is associated with one or more parameters. For example, an exponential density function,

$$f(t) = \rho e^{-\rho t}$$

where ρ is the parameter and t is the random variable. Value of ρ must be supplied externally. Now, the question is: how to calculate the optimal value? We will discuss this technique now.

A probability density function may have multiple parameters. We must find out optimal values for each of those. Of course, an optimal value will maximize the survival probability of the subject.

In general, *maximum likelihood estimation* (*MLE*) is an optimization technique for parameters of any probability distribution. Under the assumption of the probability

DOI: 10.1201/9781003255499-3

distribution suggested by MLE, the observed dataset will be most probable. It does so by maximizing the total likelihood of the observed dataset with respect to the distribution parameters. Mathematically, if $p(x, \theta)$ is the probability density function where x is the random variable and θ is the parameters, then total likelihood for n data instances is given by

$$L = \prod_{i=1}^{n} p(x_i, \theta)$$

Our objective is to maximize L and find out optimal value of θ.

As the above expression is a product of all likelihoods, maximizing this would be difficult. A better approach is to take the natural log and then maximize. This is known as log-likelihood. It is given by

$$\log L = \log \prod_{i=1}^{n} p(x_i, \theta)$$

$$= \sum_{i=1}^{n} \log p(x_i, \theta)$$

Now, the problem becomes maximizing a summation expression, which is easier. Of course, maximization is done by taking partial derivative with respect to θ, equating it to zero and solve for θ. It is given by

$$\frac{\partial(\log L)}{\partial \theta} = 0$$

If the solution of the above equation is $\hat{\theta}$, then we will use $p(x_k, \hat{\theta})$ to compute probability of an unknown data instance x_k.

MLE for Survival Function

The same approach can be taken to estimate parameter of the survival distribution. A subject observed to fail at t contributes a term $f(t, \theta)$ to the total likelihood, where f is the density function. Similarly, the subject which survives after time t contributes $S(t, \theta)$ to the total likelihood. So, full likelihood for n independent subjects is then

$$L = \prod_{i=1}^{n} [f(t, \theta)]^{c} \, [S(t, \theta)]^{1-c} \tag{3.1}$$

where c is the indicator of censoring. $c = 0$ indicates censoring and survival is evaluated at t while probability density function (PDF) gets canceled. $c = 1$ indicates occurrence of event and PDF gets evaluated while cancelation happens for survival function.

Log-likelihood expression of Equation (3.1) is given by

$$\log L = \sum c \log f(t, \theta) + \sum (1-c) \log S(t, \theta) \tag{3.2}$$

Since $f(t) = h(t)S(t)$, Equation (3.2) can be re-written as

$$\log L = \sum c \log h(t, \theta) + \sum \log S(t, \theta) \tag{3.3}$$

Solving for θ gives its MLE. We can then use this θ value to compute survival probability. We will discuss about the values of respective parameters for *Weibull*, *Gumbel* and *exponential* distribution.

We already discussed in Chapter 2 that for now all our discussion will assume that survival distribution is only dependent on time variable t. Subsequent discussions on individual probability distributions will also hold this true. We will see how survival distribution depends on other features along with time in Chapter 5.

Weibull Distribution

Weibull is one of the most popular distributions for modeling survival analysis-related problems. Its probability density function is given by

$$f(t) = \kappa\rho(\rho t)^{\kappa-1} e^{-(\rho t)^{\kappa}}$$

where ρ and κ are distribution parameters and t is the random variable denoting time. Metallic equipment of any running mechanical systems is prone to erosion, thus having a predictable failure time. *Weibull* is considered ideal for modeling these types of use cases in material science. ρ and κ are known as *scale* and *shape* parameters.

We will now plot a Weibull density with randomly generated data from a given Weibull distribution. The following python code can do that:

```
import matplotlib.pyplot as plt
import pandas as pd
from scipy.stats import weibull_min

shape = 5
scale = 0.001
weibull_samples = weibull_min.rvs(shape, 0, scale, size=1000)

pd.DataFrame(weibull_samples).plot(kind='density')
plt.legend(["Weibull"])
```

Listing 3.1 Generation of Weibull density plot.

rvs stands for *random variates*. This function generates a series of data from a Weibull distribution with a given shape and scale parameter. There is a third parameter *location(loc)* which is used for variable shifting, if needed, though it is not required in the original formulae of the PDF. That is why it is set as zero. Its output is shown below:

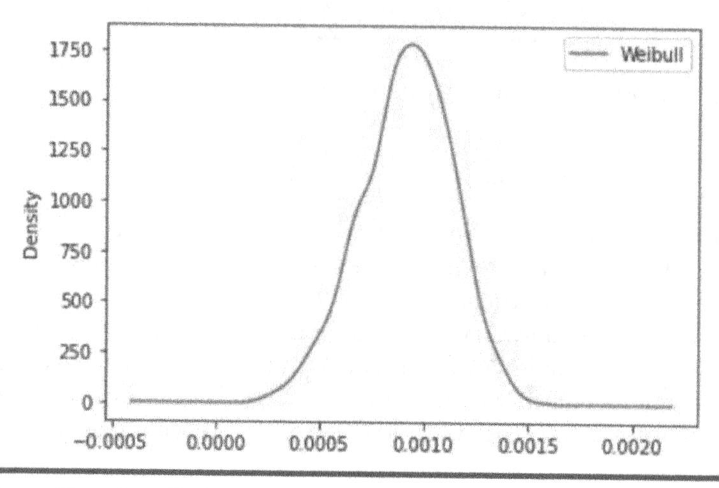

Figure 3.1 **Plot of Weibull distribution.**

Hazard and survival distributions for *Weibull* are given by

$$h(t) = \kappa\rho(\rho t)^{\kappa-1}$$

$$S(t) = e^{-(\rho t)^{\kappa}}$$

So, hazard, i.e., event/failure rate depends on time. Value of parameter κ determines the direction of dependence. For different values of κ, hazard behaves in the following ways:

1. For $\kappa = 1$, hazard becomes constant and *f(t)* becomes an *Exponential* distribution.
2. For $\kappa > 1$, hazard increases with time.
3. For $\kappa < 1$, hazard decreases with time.

MLE for ρ *and* κ

Optimal values of ρ and κ can be found by maximum likelihood estimation (MLE). Log-likelihood expression of survival function for *Weibull* is given by (obtained from Equation [3.3]),

$$\log L = \sum c \log\left[\kappa\rho(\rho t)^{\kappa-1}\right] + \sum \log\left[e^{-(\rho t)^{\kappa}}\right]$$

$$= \sum c \log\kappa + \sum c\kappa \log\rho + (\kappa - 1)\sum \log t - \rho^{\kappa} \sum t^{\kappa} \quad (3.4)$$

Estimated statistic for ρ is obtained by solving equation $\frac{\partial(\log L)}{\partial \rho} = 0$ and is given by

$$\hat{\rho} = \left(\sum \frac{c}{t^{\kappa}}\right)^{1/\kappa}$$

And for κ, solution of the Equation (3.5) is needed,

$$\sum \frac{c}{\kappa} + \sum \log t - \frac{\sum ct^{\kappa} \log t}{\sum t^{\kappa}} = 0 \tag{3.5}$$

Observe that analytically finding the solution of Equation (3.5) for κ is complex. Basically, there is no analytical closed-form solution for κ. Therefore, it is always recommended to use a numerical technique such as *Newton–Raphson* or *Nelder–Mead.* In fact, python library *lifelines* uses *Nelder–Mead* technique only.

Though detailed discussions about numerical solutions for non-linear equations like this are out of our scope, we will take a look at one of the popular methods of *Newton–Raphson* very briefly for reader's convenience.

Newton–Raphson Method for Solving MLE Equation

MLE equation can be written as $f(x) = 0$, where x is the hyper-parameter of the survival model and $f(x)$ is the partial derivative of the log-likelihood with respect to x only. This method iteratively updates x and, in each iteration, we get a better estimate as x'. This step is written as

$$x' = x - \frac{f(x)}{f'(x)}$$

where $f'(x)$ is the derivative of $f(x)$. Iteration starts with some random value of x and it stops when $x - x'$ is less than some specified threshold, i.e., there is not enough change in x between last two consecutive iterations.

For very complex functions, finding derivative analytically is also a challenge, i.e., complete algebraic form of $f'(x)$ may not be known beforehand. Hence, numerical derivative at any point is the solution. For that, we must start with two random values: x_1 and x_2 for x. Then, the next step value x_3 can be found as

$$x_3 = x_2 - f(x_2)\frac{x_2 - x_1}{f(x_2) - f(x_1)}$$

It is a variation of *Newton's* method and is known as *secant* method. Iteratively, the process goes on until we get some optimal value of x. Thus, our problem of finding solution for the non-linear MLE equation is solved.

Now, we will try to solve a practical use case using Weibull distribution. We will use *Stanford heart transplant* dataset (http://www-eio.upc.edu/~pau/cms/rdata/doc/survival/stanford2.html).

```
import pandas as pd

stanford2_df = pd.read_csv('data/stanford2.csv')
stanford2_df.head()
```

	Unnamed: 0	id	time	status	age	t5
0	139	139	86.0	1	12	1.26
1	159	159	10.0	1	13	1.49
2	181	181	60.0	0	13	NaN
3	119	119	1116.0	0	14	0.54
4	74	74	2006.0	0	15	1.26

Listing 3.2 **Reading *Standford* dataset.**

It shows survival time (*time* column) and censoring status (*status* column). Like Chapter 2, we will take a similar approach here and divide the dataset into train and test and then build the model.

```
from sklearn.model_selection import train_test_split
from lifelines import WeibullFitter

train, test = train_test_split(stanford2_df)
wbf = WeibullFitter().fit(train['time'], train['status'])
```

Listing 3.3 **Fitting Weibull model on training data of Standford dataset.**

Let us now plot hazard, survival and cumulative hazard function values over time.

```
import matplotlib.pyplot  as plt

fig, axs = plt.subplots(3, figsize=(8, 8), sharex=True)
wbf.plot_survival_function(label='survival function', ax=axs[0]).legend()
wbf.plot_hazard(label='hazard', ax=axs[1]).legend()
wbf.plot_cumulative_hazard(label='cummulative hazard', ax=axs[2]).legend()
```

Listing 3.4 **Plotting hazard functions of the fitted *Weibull* model.**

It produces the following:

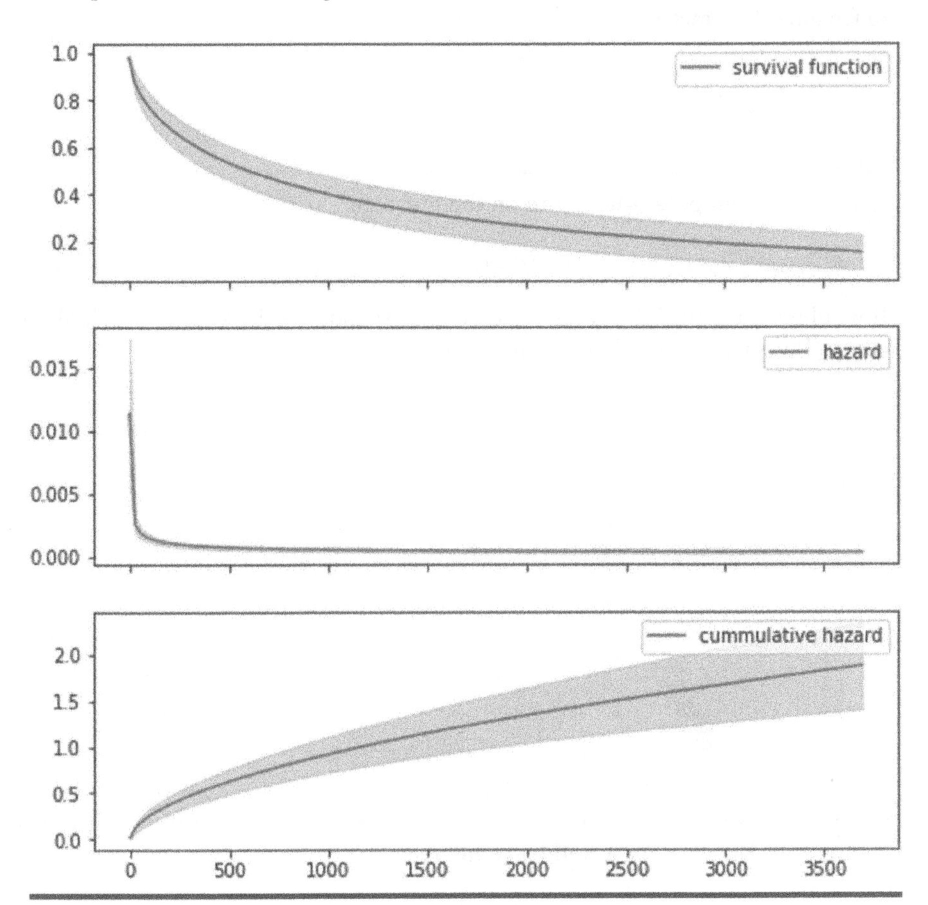

Figure 3.2 **Survival, hazard, and cumulative hazard plot.**

We can see that hazard plot is justifying expression $h(t) = \kappa\rho(\rho t)^{\kappa-1}$ for *Weibull* distribution.

Estimated parameters are

```
wbf.lambda_, wbf.rho_
```

```
(1245.6213151595903, 0.5483940149422992)
```

Listing 3.5 **Showing estimated parameters of Weibull model.**

Names may be little confusing. *lifelines* library has named κ as *rho_* and $1/\rho$ as *lambda_*. So, values are given according to that.

Now, let us test the accuracy on the test dataset. We will use *Brier Score* (refer to Chapter 2) as metric.

```
estimated_sp = wbf.predict(test['time'])
brier_score(test['status'], 1.0 - estimated_sp)

0.3642735433714426
```

Listing 3.6 **Compute Brier Score on the test data.**

Confidence Intervals of Survival Function

If we observe Figure 3.2, we can see there are expanded regions around each of the curve. These regions are shown more clearly in Figure 3.3.

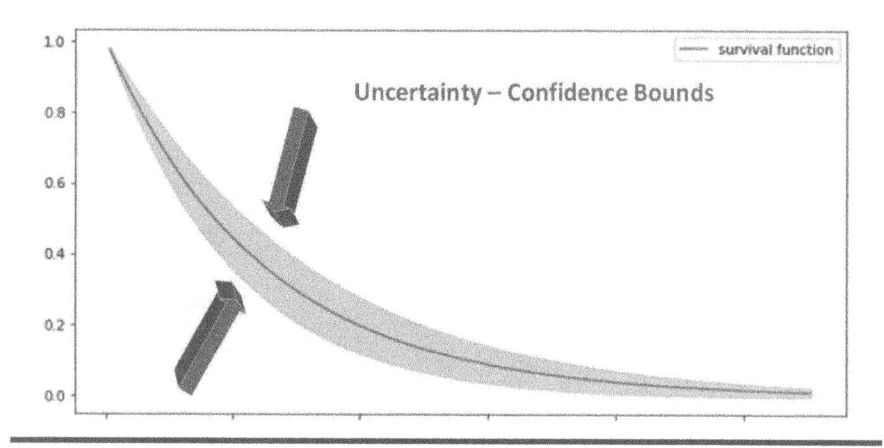

Figure 3.3 **Uncertainty–Confidence intervals.**

These indicate statistical limits of the survival function at any time instance t. Technically, these are known as *confidence intervals* (CI). It indicates the amount of uncertainty involved in the predicted survival probability. Uncertainty is related to the risk associated with the prediction. It is the potential or probable error. Now, the question is: what will we do with the risk assessment if we already know the survival probability? Risk assessment helps a lot in taking some crucial decisions, especially if the survival probability contributes to financial aspects in any business. One example could be deciding a senior citizen's life insurance premium if he/she is prone to disease. Risks associated with the survival probability of that person can be added as a weightage factor in the premium calculation.

Now, let us see how we can compute CI for *Weibull* distribution. CI for any normal distribution is easy to compute with three-sigma rule. But survival distributions (in present case, it is *Weibull*) is not normal in nature. Direct three-sigma rule is not applicable there. As a workaround, log-log transformation is applied on the survival distribution to improve its convergence to normality, i.e., the distribution becomes approximately normal. By some additional manipulation, we can get $(1-\alpha)*100\%$ upper control limits (UCL) and lower control limits (LCL) of a survival estimate $\hat{S}(t)$ as

$$\exp\left\{\log\left(\hat{S}(t)\right)\exp\left(\frac{z_{\alpha/2}}{\sqrt{n}}\right)\right\} \leq \hat{S}(t) \leq \exp\left\{\log\left(\hat{S}(t)\right)\exp\left(\frac{-z_{\alpha/2}}{\sqrt{n}}\right)\right\}$$

where value z is drawn from a standard normal distribution.

Weibull API of *lifelines* already provides CI values for all time instances. But, for our understanding of the concept, one function is implemented like below:

```
import math

def compute_confidence_intervals(train, model, estimate_name):
    nu = len(train[train.status == 1])
    lower_col = estimate_name + '_lower'
    upper_col = estimate_name + '_upper'
    ci_limits = pd.DataFrame(columns=['time', lower_col,upper_col])
    z = 1.96
    for t_i in model.survival_function_.index:
        surv_prob = wbf.survival_function_.loc[t_i,estimate_name]
        ci_limit = {}
        ci_limit['time'] = t_i
        ci_limit[lower_col] = (math.exp(math.log(surv_prob) * math.exp(z/math.sqrt(nu))))
        ci_limit[upper_col] = (math.exp(math.log(surv_prob) * math.exp(-z/math.sqrt(nu))))
        ci_limits = ci_limits.append(ci_limit, ignore_index=True)

    return ci_limits
```

Listing 3.7 Compute CI limits for *Weibull* survival distribution.

We can see that z is kept at 1.96, which is the standard normal distribution value for 95% level of significance. Now, this function can be used over training dataset like below:

```
compute_confidence_intervals(train, wbf, 'Weibull_estimate').head()
```

Listing 3.8 Call *compute_confidence_interval* function on training dataset.

And it produces the following result:

	time	Weibull_estimate_lower	Weibull_estimate_upper
0	1.000000	0.974746	0.983339
1	27.963504	0.853042	0.900858
2	54.927007	0.794398	0.859682
3	81.890511	0.750861	0.828436
4	108.854015	0.715379	0.802507

Figure 3.4 Result showing CI limits for various time instances.

Gumbel Distribution

Sometimes, log transformation on time scale is needed if time differences of events are very large. From data science practice, it is known that log transformation helps to mitigate effect of outliers. Same can be applied here also. But it influences the distribution. Log transformation converts the *Weibull* density to *Gumbel* density. Before going into that detail, we need to understand the effect of variable transformations, in general, on integrals.

Transformation of Variables for Integrals – Jacobian

Consider a double integral,

$$\iint f(x,y)\ dx\,dy.$$

Suppose we want to replace variables x and y by u and v respectively. So, $f(x,y)$ will be transformed into some function $g(u,v)$. The *Jacobian* of this change of variables is defined as the determinant of the partial derivative matrix like below:

$$J = \begin{vmatrix} \frac{\partial x}{\partial u} & \frac{\partial x}{\partial v} \\ \frac{\partial y}{\partial u} & \frac{\partial y}{\partial v} \end{vmatrix}$$

We know that determinant is the factor by which area changes due to transformation of variables. As any integral is nothing but an area so, the integral mentioned above can be re-written as

$$\iint f(x,y)\ dx\,dy = \iint g(u,v)\ J\ du\,dv$$

The above relation is true for single well as multi-variables case. This concept is used to derive Gumbel distribution.

Inception of Gumbel Distribution

Let us consider transformation variable u as $u = \log t$, i.e., log transformation of time scale of survival dataset. If there are two random variables u and t, where $p_u(u)$ and $p_t(t)$ are their respective probability density functions, then from the concept of variable transformation effect on integrals, we can say

$$J = \left|\frac{dt}{du}\right|$$

As it is single variable case, partial derivative can be replaced by full derivative. Basically, probability density functions are nothing but integrals and the probability is the area under them. So,

$$p_u(u) = p_t(t)\left|\frac{dt}{du}\right| \tag{3.6}$$

As $u = \log t$, then $t = e^u$ and $\frac{dt}{du} = e^u$.

Now, we will apply this transformation on PDF of *Weibull*, i.e.,

$$p_t^{Weibull}(t) = \kappa\rho(\rho t)^{\kappa-1}\ e^{-(\rho t)^{\kappa}}.$$

Then using Equation (3.6), we get

$$p_u(u) = \kappa\rho(\rho t)^{\kappa-1}\ e^{-(\rho t)^{\kappa}}\ e^u$$

$$= \kappa\rho(\rho e^u)^{\kappa-1}\ e^{-(\rho e^u)^{\kappa}}\ e^u\ (\text{replacing } t \text{ with } e^u)$$

Rearranging the expression, we get

$$p_u^{Gumbel}(u) = \kappa\rho^{\kappa} e^{\kappa u - \rho^{\kappa} e^{\kappa u}} \tag{3.7}$$

Equation (3.7) is the probability density function of *Gumbel* distribution with random variable u which is log-transformed time. It is a distribution of maximum number (or minimum) among set of samples of different distributions. It follows the family of *extreme value distributions.*

We will now plot a Gumbel density with randomly generated samples like we did for Weibull. The following python code can do that:

```
import numpy as np

k, rho = 0.6, 0.1
gumbel_samples = np.random.gumbel(k, rho, 1000)

pd.DataFrame(gumbel_samples).plot(kind='density')
plt.legend(["Gumbel"])
```

Listing 3.9 **Generation of Gumbel density plot.**

And it produces the following graph:

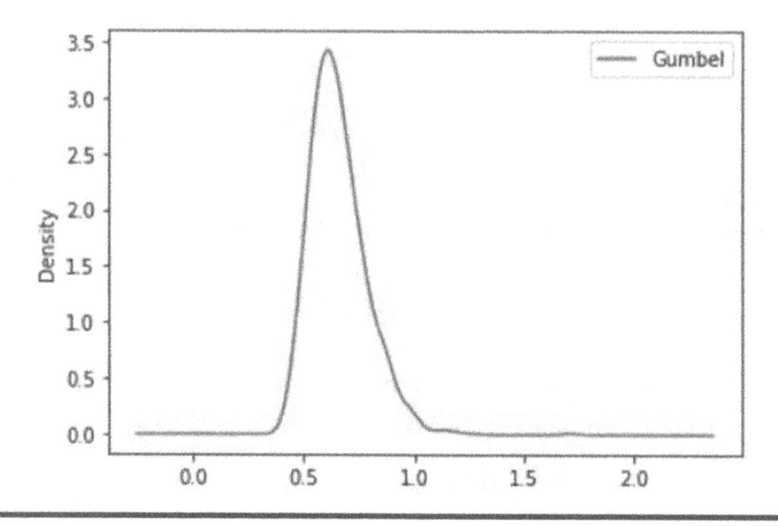

Figure 3.5 **Plot of Gumbel distribution.**

Survival and Hazard Function of Gumbel Distribution

Survival function of *Gumbel* is given by

$$S(u) = e^{-\rho^{\kappa} e^{\kappa u}}$$

And hazard function is given by

$$h(u) = \kappa \rho^{\kappa} e^{\kappa u}$$

Gumbel distribution is often called *log-Weibull* distribution. As mentioned earlier, the distribution of maximum number among a set of numbers follows Gumbel distribution. If these numbers indicate measurement, then distribution of maximum numbers can capture extreme values. Hence, *Gumbel* distribution is useful in modeling any sort of unnatural or extreme behavior (especially modeling natural calamities such as extreme earthquake, flood, etc.). In simple terms it is useful for "*rare events*".

If *Weibull* modeling does not give good accuracy for rare events, then we can log-transform the time variable and use *Gumbel* distribution to model a use case for survival analysis.

Exponential Distribution

We already saw one example of *exponential* density in Chapter 2. Here we will discuss it in more detail. Density, survival and hazard functions are given by

$$f(t) = \rho e^{-\rho t}$$

$$S(t) = e^{-\rho t}$$

$$h(t) = \rho$$

Hazard is a constant for *exponential* distribution. It reflects the property of the distribution reasonably called *lack of memory*. It is ideal for situations where event rate does not change over time.

MLE for ρ

Exponential density has only parameter ρ. Like *Weibull*, best value for ρ can be estimated by ML. Log-likelihood expression obtained from Equation (3.3) is given by

$$\log L = \sum \text{c} \log \rho + \sum \log\left[e^{-\rho t}\right]$$

$$= \sum \text{c} \log \rho - \rho \sum \log t$$

Estimated statistic for ρ is obtained by solving equation $\frac{\partial(\log L)}{\partial \rho} = 0$ and is given by

$$\rho = \sum \frac{c}{t}$$

c is as usual a censoring indicator. It takes values 0 or 1 at different time instance t.

We will now model the same *Stanford heart transplant* problem with *exponential* distribution.

```
from lifelines import ExponentialFitter

exf = ExponentialFitter().fit(train['time'], train['status'])

fig, axs = plt.subplots(3, figsize=(8, 8), sharex=True)
exf.plot_survival_function(label='survival function', ax=axs[0]).legend()
exf.plot_hazard(label='hazard', ax=axs[1]).legend()
exf.plot_cumulative_hazard(label='cummulative hazard', ax=axs[2]).legend()
```

Listing 3.10 **Fit exponential model and plot hazard function.**

It produces the following graph:

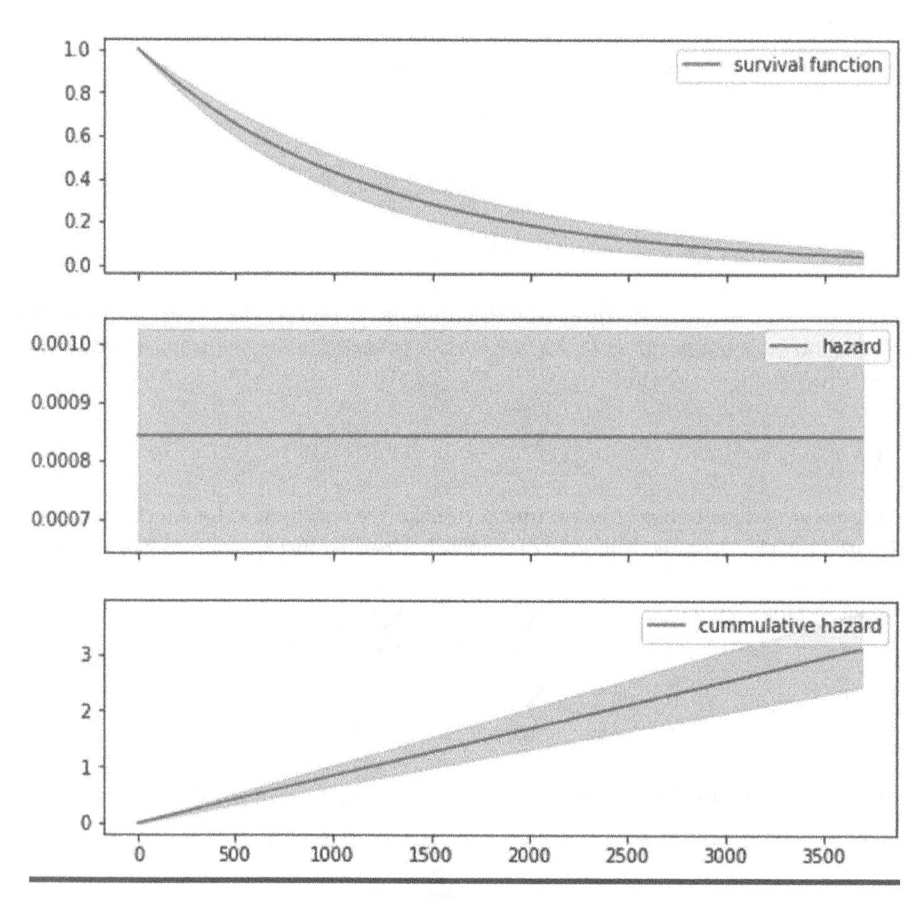

Figure 3.6 **Survival, hazard, cumulative hazard plot for *exponential*.**

Value of the distribution parameter is

```
exf.lambda_

1187.5964855626016
```

Listing 3.11 **Show parameters of *exponential* model.**

lifelines library uses name *lamda_* for variable ρ. Value of *lamda_* is 1/ρ.

Comparison of Models

We saw how we can use *Brier Score* to get the accuracy of any model. It can be used for comparing models coming from different distributions. Like any other accuracy metric, *Brier Score* is preferable to use on test dataset or cross validation set. But there is a common problem like any other traditional machine learning model. It is *training time* for very large dataset. For survival models, mostly MLE is considered as training. Few of them rely on complex iterative numerical techniques for solving the MLE equation. Cross validation technique while training will be quite time-consuming in that case. We will also discuss some other challenges regarding this in Chapter 5 while dealing with covariates. But for now, the challenge is time. To solve this, an alternative metric is chosen which can be computed on the training data while doing the training itself. We will discuss it next.

Akaike Information Criterion (AIC)

Akaike information criterion (AIC) is given by the expression

$$\text{AIC} = -2\log L + 2k$$

where log L is the log-likelihood of the model and k is the number of parameters. Log-likelihood is computed on the training dataset while building the model. Parameter k denotes number of parameters of the underlying probability distribution family. Low AIC score indicates a better model. Now, let us compare our two models, *Weibull* and *exponential* on the entire training dataset.

```
wbf.AIC_, exf.AIC_
```

```
(1283.7163423173122, 1343.2279920982928)
```

Listing 3.12 Show AIC values of *exponential* and *Weibull* model.

Weibull is a better model designed on the training dataset as it has lower AIC. AIC penalizes complex models with higher number of parameters. A simpler model with fewer parameter is always preferable over any other one where difference in log-likelihood is negligible. AIC score abides by this principle and gives results accordingly.

Chapter 4

Non-Parametric Models

We discussed various statistical and parametric approaches of modeling survival distributions in Chapter 3. But in most of the cases, especially with human or animal survival or some bio statistics use case, it is hard to know parametric probability distributions beforehand. In this chapter, we will discuss non-parametric estimations of survival distributions. Our discussion will start with *Kaplan–Meier* (KM) estimator.

Kaplan–Meier Estimator

In parametric approaches, we saw how we assumed one mathematical form of the distribution and tried to fit that into our dataset. Here the advantage is that we know the formulae of the density and distribution function and the maximum likelihood estimation (MLE) can guide us to find out optimal values of parameters. On the other hand, non-parametric models are known to be distribution-free. We cannot assume any concreate mathematical formulae for the probability density. Probability computation must be done here intuitively. *Kaplan–Meier* estimator is one of the most widely used non-parametric techniques for modeling survival distributions. Let us now discuss its details.

Suppose n_i is the population size at risk at time instance t_i and d_i is the number of events occurred at t_i, then survival probability function is estimated by

$$S_{KM}(t) = \prod_{t_i < t} \left(1 - \frac{d_i}{n_i}\right)$$

DOI: 10.1201/9781003255499-4

d_i and n_i are simple counts from the population. $\frac{d_i}{n_i}$ is the event or failure probability at time instance t_i. n_i is cumulative in nature and keeps on getting updated over the time as the number of events affects total population. Following relationship holds true if the event is a death event and it reduces the population size,

$$n_{i+1} = n_i - d_i$$

Let us consider *Stanford2* dataset again.

```
import pandas as pd

stanford2_df = pd.read_csv('data/stanford2.csv')
stanford2_df.head()
```

	Unnamed: 0	id	time	status	age	t5
0	139	139	86.0	1	12	1.26
1	159	159	10.0	1	13	1.49
2	181	181	60.0	0	13	NaN
3	119	119	1116.0	0	14	0.54
4	74	74	2006.0	0	15	1.26

Listing 4.1 **Read *Standford* dataset.**

None of the records is repeated over here and the total risk set size is 184, i.e., the total of number of records. Let us now sort the records by *time* column and check how many deaths happened in each time instance.

```
stanford2_df_sorted = stanford2_df.sort_values(by=['time'])
event_counts = stanford2_df_sorted[stanford2_df_sorted['status'] == 1].groupby('time').size().reset_index(name='event/death')

event_counts.plot(x='time',y='event/death')
```

Listing 4.2 **Group by *Standford* dataset by *time* and plot time vs event.**

We will a get a graph like below:

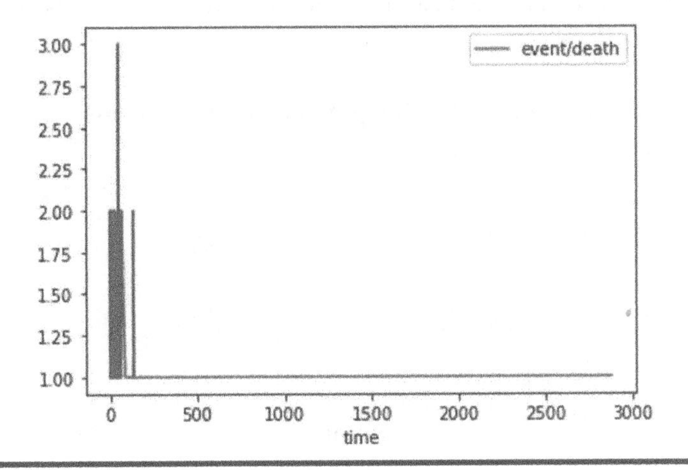

Figure 4.1 **Count plot of time vs event for *Standford* dataset.**

It shows that for few time instances, there are multiple deaths, and hence $d_i > 1$. Let us have a look at the dataset sorted by time.

```
stanford2_df_sorted
```

	Unnamed: 0	id	time	status	age	t5
71	21	21	0.5	1	41	0.87
171	16	16	1.0	1	54	0.47
22	184	184	1.0	0	27	NaN
15	133	133	1.0	1	21	0.47
60	183	183	2.0	0	39	NaN
...	...	...	...	...	...	...
141	25	25	2878.0	1	49	0.75
39	38	38	2984.0	0	32	0.19
59	36	36	3021.0	0	38	0.98
101	26	26	3410.0	0	45	0.98
70	19	19	3695.0	0	40	0.38

184 rows × 6 columns

Listing 4.3 **Show *Standford* dataset records sorted by *time*.**

We can count d_i, keep on updating n_i and compute the KM estimator for each time entry. Step-wise values for first three time instances will be like below:

Table 4.1 **Step-Wise Computation of KM Estimator**

t_i	d_i	n_i	d_i/n_i	$S_{KM}(t)$
0.5	1	184	1/184	$\left(1-\frac{1}{184}\right)$
1.0	2	183	2/183	$\left(1-\frac{1}{184}\right)\times\left(1-\frac{2}{183}\right)$
2.0	1	181	1/181	$\left(1-\frac{1}{184}\right)\times\left(1-\frac{2}{183}\right)\times\left(1-\frac{1}{181}\right)$

We can observe that n_i is getting updated over time and is decreasing. $S_{KM}(t)$ is as usual the survival probability at time t.

Derivation of $S_{KM}(t)$

Expression of $S_{KM}(t)$ can be derived from the MLE process like other models. This discussion will be core mathematical in nature. Readers who are not interested may skip this.

From the probability computation discussion with *veteran* dataset, we can see that time should be considered discrete rather than continuous for KM estimator (at least to evaluate its expression). If discreate hazard rate is h_i, then survival function can be defined as

$$S_{KM}(t) = \prod_{t_i<t}(1-h_i)$$

Likelihood expression is given by (from our discussion in Chapter 3)

$$L = \prod_{t_i<t} h_i^{d_i}(1-h_i)^{n_i-d_i}$$

(As there are d_i number of events)

We must find out an optimal $\hat{h}_i$ for which L has a maximum value. Log-likelihood expression is now given by

$$\log L = \sum \left[d_i \log h_i + (n_i - d_i) \log(1 - h_i) \right]$$

Now, $\frac{\partial \log L}{\partial h}$ is given by

$$\frac{\partial \log L}{\partial h} = \frac{d_i}{h_i} - \frac{n_i - d_i}{1 - h_i}$$

Equating $\frac{\partial \log L}{\partial h} = 0$ and solving for h, we get the optimal value $\hat{h}$,

$$\hat{h} = \frac{d_i}{n_i}$$

Putting $\hat{h}$ in $S_{KM}(t)$, we get the final expression which we just used for probability computation for *veteran* dataset.

$$S_{KM}(t) = \prod_{t_i < t} \left(1 - \frac{d_i}{n_i} \right)$$

Now, we will use Python *lifelines* library to model the KM estimator with veteran dataset.

```
from lifelines import KaplanMeierFitter

kmft = KaplanMeierFitter()
kmft.fit(stanford2_df_sorted['time'],stanford2_df_sorted['status'])
```

Listing 4.4 Fit KM model for Standford dataset with *lifelines* API.

Let us see the survival function plot.

```
kmft.plot_survival_function()
```

Listing 4.5 Plot survival function of the fitted KM model.

And it produces the following plot:

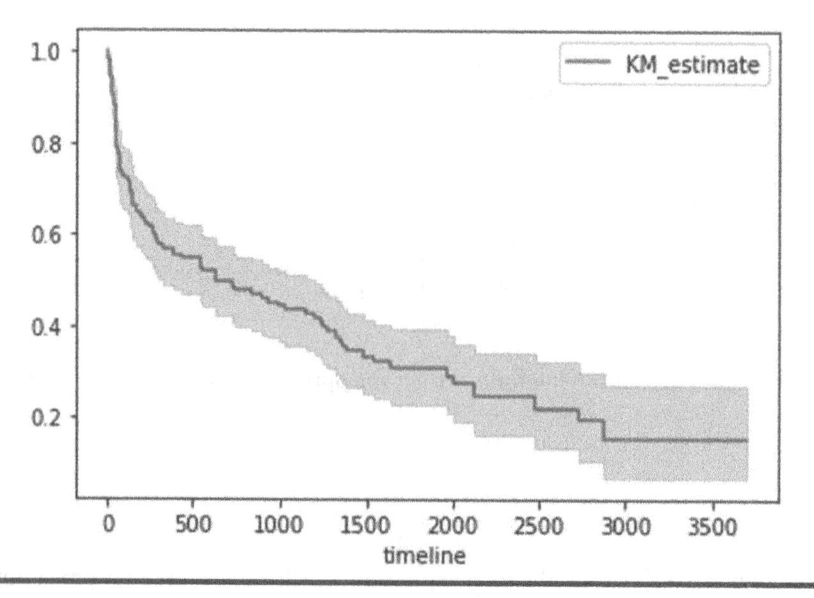

Figure 4.2 **Survival function plot of fitted KM model.**

The curve looks exactly like what was depicted in Chapter 2. Due to its discrete nature, KM estimate curves always produce step-wise looks.

Computation of Survival Function for Unknown Time Instance

Now, we will get the survival probabilities for three different time instances using the built model.

```
pd.DataFrame(kmft.survival_function_at_times(times=[1.5,2878.0,3500.0]))
```

	KM_estimate
1.5	0.983696
2878.0	0.154919
3500.0	0.154919

Listing 4.6 **Get survival probabilities for three time entries using fitted model.**

If we have a look at the dataset, we will notice that time instance 2878.0 is present there, but 1.5 and 3500.0 are new data points. It is a quite normal case and not a big deal for a parametric model. For example, had it been modeled using *Weibull* distribution, we could easily get the survival probability for any time as we have the formulae for the probability density function and survival function. Let us see this in detail. Survival function for *Weibull* distribution

$$S_{Weibull}\left(t\right) = e^{-(\rho t)^{\kappa}}$$

We know the estimates for ρ and κ are $\hat{\rho}$ and $\widehat{\kappa}$, respectively. So, the survival probability at 3500.0 would be

$$S_{Weibull}\left(3500.0\right) = e^{-(3500.0\hat{\rho})^{\hat{\kappa}}}$$

It is easy to compute. But for a non-parametric model, it is not. There is an event occurred at 2878.0, and hence n_i and d_i can be obtained for that time instance and KM estimates can be computed. But what about 1.5 and 3500.0? There may or may not be any event and somehow that was not captured in the dataset. To compute Kaplan–Meier estimate, we must know n_i and d_i; otherwise by theory, computation is not possible. Unlike *Weibull*, it does not provide a concrete algebraic function of time t for survival probability. But still, we can see that *KaplanMeierFitter* from *lifelines* library can provide us the probabilities for these unknown timestamps. How is that possible? *Linear interpolation* is the answer. *Interpolation* is one of the very common estimation techniques. For our convenience, we will discuss it in very short in regard to KM estimation.

If there are any two existing time instances t_a and t_b, corresponding available survival probabilities S_a and S_b, then survival probability S_x for any unknown time instance t_x can be found by linear interpolation like below:

$$S_x = S_a + \left(S_b - S_a\right)\frac{t_x - t_a}{t_b - t_a}$$

where $t_a < t_x < t_b$.

When we have a collection of unknown time instances for which survival probabilities are to be found out, we can do a series of binary search to find out upper and lower bound (t_b and t_a, respectively) pair for each time instance from the existing time values. Thus, we have a collection of linear interpolation models like above for each time bound pair. Our job is done then. Using the models, we can find out corresponding S_x values.

Confidence Intervals of the Survival Function – Greenwood's Estimator

We already discussed about concept and importance of confidence intervals in Chapter 3. For *Kaplan–Meier* model, one technique known as *Greenwood's formula* is used to compute confidence intervals. Before that, we need a basic idea of Taylor series approximated as explained next.

Taylor series expansion says that a function $f(x)$ can be approximated about point $x = a$ with the expansion like below:

$$f(x) = f(a) + f'(a)(x-a) + \frac{f''(a)}{2!}(x-a)^2 + \ldots + \frac{f^n(a)}{n!}(x-a)^n + .$$

where $f'(x) = \frac{\partial f(x)}{\partial x}$ and $f'(a) = \left.\frac{\partial f(x)}{\partial x}\right|_{x=a}$.

For our convenience, we will consider only first two terms of the series,

$$f(x) \cong f(a) + f'(a)(x-a)$$

As variance is only affected by scale and not by origin, an estimation of variance of $f(x)$ would be

$$\widehat{\text{var}}[f(x)] = [f'(a)]^2 \ \text{var}(x-a)$$

Now, to derive CI, we can start by finding an approximation of log x which will become useful later. Here, $f(x) = \log x$ and $a = \mu$ (mean of x), so using Taylor series,

$$\log x = \log\mu + (x-\mu)\frac{1}{\mu} \left(\text{As } f'(x) = \frac{1}{x}, \text{ hence } f'(\mu) = \frac{1}{\mu} \right)$$

And we get the estimation of variance of log x,

$$\widehat{\text{var}}[\log x] = [f'(\mu)]^2 \ \text{var}(x-\mu)$$

$$= \frac{\widehat{\sigma^2}}{\mu^2}$$

where $\text{var}(x-\mu) = \widehat{\sigma^2}$).

Taking the logarithm of $S_{KM}(t)$, we get

$$\log\left[S_{KM}(t)\right] = \sum \log\left(1 - \frac{d_i}{n_i}\right)$$

$$= \sum \log \widehat{p_i}$$

where $\widehat{p_i} = 1 - \frac{d_i}{n_i}$).

One interesting point to be noted here, number of events d_i follows a Binomial distribution with parameters (n_i, p_i) and its variance would be $n_i p_i (1 - p_i)$. A good estimate of p_i is $\widehat{p_i}$ as given above. We therefore have

$$\text{var}\left(\widehat{p_i}\right) = \text{var}\left(1 - \frac{d_i}{n_i}\right)$$

$$= \frac{\text{var}(d_i)}{n_i^2}$$

$$= \frac{n_i p_i (1 - p_i)}{n_i^2}$$

$$\cong \frac{\widehat{p_i}\left(1 - \widehat{p_i}\right)}{n_i}$$

From the expression of $\widehat{\text{var}}\left[\log x\right]$, we can get an estimate of $\text{var}\left[\log \widehat{p_i}\right]$ in a similar manner.

$$\widehat{\text{var}}\left[\log \widehat{p_i}\right] = \frac{\text{var}\left(\widehat{p_i}\right)}{\left[E\left(\widehat{p_i}\right)\right]^2}$$

$$= \frac{\widehat{p_i}\left(1 - \widehat{p_i}\right)}{n_i \widehat{p_i}^2}$$

$$= \frac{1 - \widehat{p_i}}{n_i \widehat{p_i}}$$

So,

$$\widehat{\text{var}}\left[\sum \log \widehat{p_i}\right] = \sum \frac{1-\widehat{p_i}}{n_i \widehat{p_i}}$$

And from expansion of Taylor series, we get

$$\text{var}\left[\log\left[\widehat{S_{KM}}(t)\right]\right] = \frac{\text{var}\left[\widehat{S_{KM}}(t)\right]}{\left[\widehat{S_{KM}}(t)\right]^2}$$

As we know, $\text{var}\left[\log\left[S_{KM}(t)\right]\right] = \widehat{\text{var}}\left[\Sigma \log \widehat{p_i}\right]$, then

$$\text{var}\left[\widehat{S_{KM}}(t)\right] = \left[\widehat{S_{KM}}(t)\right]^2 \sum \frac{1-\widehat{p_i}}{n_i \widehat{p_i}}$$

Standard error of $S_{KM}(t)$ is then given by

$$s\left[\widehat{S_{KM}}(t)\right] = \widehat{S_{KM}}(t)\sqrt{\sum \frac{1-\widehat{p_i}}{n_i \widehat{p_i}}}$$

This is known as *Greenwood's* estimate of *standard error.*

If $S_{KM}(t)$ follows normal distribution, then we can find out confidence intervals by ranges as given by

$$\text{CI}\left[\widehat{S_{KM}}(t)\right] = \widehat{S_{KM}}(t) \pm z_{\alpha/2} s\left[\widehat{S_{KM}}(t)\right]$$

where z follows standard normal distribution $N(0,1)$ and α is level of significance.

But there is a small problem. As it is a non-parametric technique, we do not really know whether $S_{KM}(t)$ follows normal distribution or not. In fact, it does not, and hence confidence interval estimates may not be fully accurate. One high level estimate of normal density variable can be found by log transformation on non-normal or skewed normal variable. Another special approach is log-log transformation, which is generally used for Greenwood's estimator. Transformed variable now becomes $\log\left[-\log\left[\widehat{S_{KM}}(t)\right]\right]$. Like previous computations so far, we can have

$$\text{var}\left[-\log\left[\widehat{S_{KM}}(t)\right]\right] = \sum \frac{1-\widehat{p_i}}{n_i \widehat{p_i}}$$

As var(x) = var(–x)

Using the expression for var[log x], we can get estimate for $\text{var}\left[\log\left[-\log\left[\widehat{S_{KM}}(t)\right]\right]\right]$,

$$\text{var}\left[\log\left[-\log\left[\widehat{S_{KM}}(t)\right]\right]\right] = \frac{1}{\left[\log\left[\widehat{S_{KM}}(t)\right]\right]^2}\sum\frac{1-\widehat{p_i}}{n_i\,\widehat{p_i}}$$

So, the standard error of $\log\left[-\log\left[\widehat{S_{KM}}(t)\right]\right]$ would be

$$s\left[\log\left[-\log\left[\widehat{S_{KM}}(t)\right]\right]\right] = \frac{1}{\log\left[\widehat{S_{KM}}(t)\right]}\sqrt{\sum\frac{1-\widehat{p_i}}{n_i\,\widehat{p_i}}}$$

Assuming that $\log\left[-\log\left[\widehat{S_{KM}}(t)\right]\right]$ is normally distributed, confidence intervals of $\log\left[-\log\left[\widehat{S_{KM}}(t)\right]\right]$ is given by

$$\text{CI}\left[\log\left[-\log\left[\widehat{S_{KM}}(t)\right]\right]\right] = \log\left[-\log\left[\widehat{S_{KM}}(t)\right]\right] \pm z_{\alpha/2}\,s\left[\log\left[-\log\left[\widehat{S_{KM}}(t)\right]\right]\right]$$

As this interval is not for the original variable $\widehat{S_{KM}}(t)$ rather for the log-log transformed scale, we need to transform it back to the original scale. The first transformation to the exponent scale will give CI limits for $\log\left[\widehat{S_{KM}}(t)\right]$ like below:

$$\text{CI}\left[\log\left[\widehat{S_{KM}}(t)\right]\right] = \log\left[\widehat{S_{KM}}(t)\right]e^{\pm z_{\alpha/2}s\left[\log\left[-\log\left[\widehat{S_{KM}}(t)\right]\right]\right]}$$

And the second transformation to the exponent scale will give CI limits of our original function $\widehat{S_{KM}}(t)$

$$\text{CI}\left[\widehat{S_{KM}}(t)\right] = \left[\widehat{S_{KM}}(t)\right]^{e^{\pm z_{\alpha/2}s\left[\log\left[-\log\left[\widehat{SKM}(t)\right]\right]\right]}}$$

which can be decomposed as CI limit pair $\left(\left[\widehat{S_{KM}}(t)\right]^{e^{z_{\alpha/2}s\left[\log\left[-\log\left[\widehat{SKM}(t)\right]\right]\right]}}, \left[\widehat{S_{KM}}(t)\right]^{e^{-z_{\alpha/2}s\left[\log\left[-\log\left[\widehat{SKM}(t)\right]\right]\right]}}\right)$.

We can get the CI limit values for our fitted survival function like below:

```
kmft.confidence_interval_survival_function_
```

	KM_estimate_lower_0.95	KM_estimate_upper_0.95
0.0	1.000000	1.000000
0.5	0.962052	0.999233
1.0	0.950307	0.994712
2.0	0.950307	0.994712
3.0	0.942962	0.991762
...	...	...
2878.0	0.069245	0.271866
2984.0	0.069245	0.271866
3021.0	0.069245	0.271866
3410.0	0.069245	0.271866
3695.0	0.069245	0.271866

Listing 4.7 **Confidence interval of survival probabilities.**

It uses Greenwood's estimator only.

Log-Rank Test

Suppose one medical drug test is being performed on two groups of people and their survival function must be compared for this. Groups may represent two treatment or one control and one treatment group. Survival curves as usual will indicate trend of survival probabilities. Comparison of the same between two groups will tell us whether these two curves are equal in *overall sense* or not. For KM curves, this test of equality is vital as we do not have any predefined distribution. We cannot test based on some statistical assumption about the distribution. For that, we need some sort of significance test which will statistically verify the equality.

Let us have a look at the generated survival curves for two treatment groups. For this, we will use *veteran* dataset as discussed in earlier chapters. We saw that this dataset contains records for two treatment groups (1 and 2). We will break it into two parts-each for one group.

```
veteran_df_1 = veteran_df[veteran_df.trt==1].drop(['trt','Unnamed: 0'], axis=1)
veteran_df_2 = veteran_df[veteran_df.trt==2].drop(['trt','Unnamed: 0'], axis=1)
```

Listing 4.8 **Prepare veteran dataset (break into two treatment groups).**

Then, we can fit two KM curves and visualize it like below:

```
from lifelines import KaplanMeierFitter

kmft_1 = KaplanMeierFitter()
kmft_1.fit(veteran_df_1['time'],veteran_df_1['status'], label='Treatment Group 1')

kmft_2 = KaplanMeierFitter()
kmft_2.fit(veteran_df_2['time'],veteran_df_2['status'], label='Treatment Group 2')

kmft_1.plot_survival_function()
kmft_2.plot_survival_function()
```

Listing 4.9 **Fit KM models for two different treatment groups.**

It will produce the following curves:

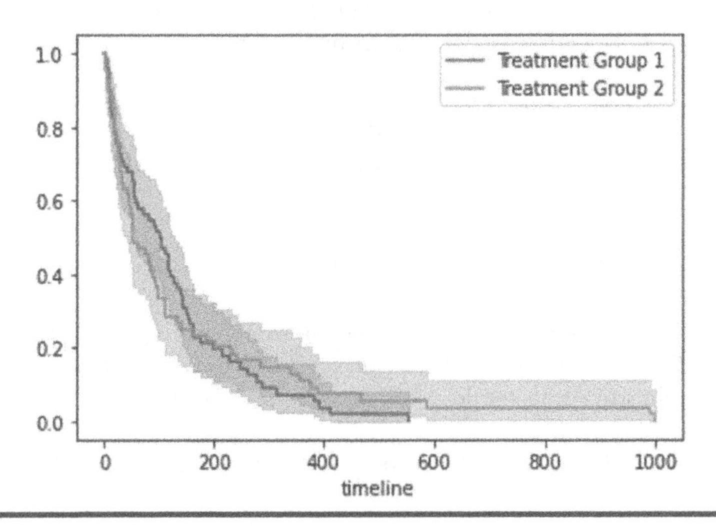

Figure 4.3 **Plot of two different KM curves.**

Both curves have some overlapping with each other. But just by looking at them, it is never possible to conclude that whether they represent same distribution or not. Log-rank test is a large sample chi-square (χ^2) test that uses a chi-square test statistic to compare two or more KM curves. Before going into detail, let us have a short information about a generic χ^2 test.

A χ^2 distribution is a probability distribution of the sum of squares of a collection of standard normal random variables. This random variable is given by

$$Q = \sum_{i=1}^{k} Z_i^2$$

where Z_i are individual standard normal random variables. This Q follows χ^2 distribution with k degrees of freedom, i.e.,

$$Q \sim \chi^2\ (k)$$

It can be proved that a Binomial distribution (n, p, q) can be approximated by a normal distribution with the random variable,

$$\chi = \frac{m - np}{\sqrt{npq}}$$

where m is observed number of successes in n trials. Of course, np is the expected number of successes. Squaring, rearranging, and summing χ, a new cumulative statistic χ^2 can be derived and its expression is given by

$$\chi^2 = \sum_{i=1}^{n} \frac{(O_i - E_i)^2}{E_i}$$

where O_i is the observed number of occurrences and E_i is the expected number of occurrences.

Now, the interesting part comes. How will we apply this χ^2 test statistic in comparing KM curves? From the discussion of *Greenwood's* estimator, we know that d_i, i.e., number of events (or deaths) follows Binomial distribution. So, that can be approximated as χ^2 distribution.

There are different types of log-rank tests available. We will use the one as,

$$(LogRank)_i = \frac{\sum_{j=1}^{t} (O_{ij} - E_{ij})^2}{\text{var}(O_i - E_i)}$$

where i is group number and t is number of time instances present in the dataset. It is known as ***Mantel–Haenszel logrank test.***

It follows a χ^2 distribution with 1 degree of freedom. As per the convention, we need to compute log-rank test statistic for any one of the treatment groups data (i = 1 or i = 2). The null hypothesis H_0 says KM curves of two groups are identical and the alternative H_1 says they are not. As usual a high p-value > 0.05 accepts H_0; otherwise H_1 is accepted. Let us now see how to compute O_i and E_i for KM curves.

E_{ij} is the expected number of events for group i at time instance t_j. Its expression is given by

$$E_{ij} = \frac{n_{ij} \sum_i d_{ij}}{\sum_i n_{ij}}$$

where n_{ij} is the total number of records at risk at time instance t_j for group i, d_{ij} is the number of events at time instance t_j for group i, $\sum_i n_{ij}$ is the total number of records at risk at time instance t_j across all groups, and $\sum_i d_{ij}$ is the total number of events at time instance t_j across all groups.

Let us now compute one expected number of events E_{13}, i.e., expected number of events E at time instance t_3 for treatment group 1 (for *veteran* dataset). Its expression would be

$$E_{13} = \frac{n_{13}\left(d_{13} + d_{23}\right)}{n_{13} + n_{23}}$$

For this, we should look at the event status by sorting the time values.

```
veteran_df_sorted = veteran_df.sort_values(by=['time'])
veteran_df_sorted
```

	Unnamed: 0	trt	celltype	time	status	karno	diagtime	age	prior
76	77	2	squamous	1	1	20	21	65	10
84	85	2	squamous	1	1	50	7	35	0
94	95	2	smallcell	2	1	40	36	44	10
52	53	1	adeno	3	1	30	3	43	0
17	18	1	smallcell	4	1	40	2	35	0
...	...	...	...	...	...	...	...	...	...
82	83	2	squamous	467	1	90	2	64	0
57	58	1	large	553	1	70	2	47	0
77	78	2	squamous	587	1	60	3	58	0
74	75	2	squamous	991	1	70	7	50	10
69	70	2	squamous	999	1	90	12	54	10

Listing 4.10 **Sort *veteran* dataset by time.**

Record number 52 represents time instance t_3 because 76 and 84 should be clubbed together as records for t_1. At the beginning, i.e., at time t_1, population values (n_{11} and n_{21}) for two groups are obtained like below:

```
n11 = len(veteran_df_1)
n21 = len(veteran_df_2)
n11, n21

(69, 68)
```

Listing 4.11 Get sizes of two different treatment group records.

From the dataset, we can identify that $d_{11} = 0$ and $d_{21} = 2$, i.e., number of events at time t_1 for treatment group 1 and 2, respectively. In a similar fashion, $d_{13} = 1$ and $d_{23} = 0$. From the relation, $n_{i+1} = n_i - d_i$ (as discussed in earlier section), we get $n_{13} = 68$ and $n_{23} = 65$. So, now E_{13} is evaluated as

$$E_{13} = \frac{68(1+0)}{68+65} \cong 0.51$$

O_{13} is nothing but d_{13}. So, $O_{13} - E_{13} = 0.49$. Like this, we can compute the numerator part of the log-rank statistic, i.e., $\sum_{j=1}^{n}(O_{ij} - E_{ij})^2$. Here, $n = 137$, i.e., total number of time instances captured in the dataset. Denominator of log-rank is a variance metric. It is the variance of the differences between observed and expected number of events. For more than two groups, formulae of variance would be very complex and involve covariance computations. But for two groups, it can be given by

$$\text{var}(O_i - E_i) = \sum_{j=1}^{t} \frac{n_{1j} n_{2j} \left(d_{1j} + d_{2j}\right)\left(n_{1j} + n_{2j} - d_{1j} - d_{2j}\right)}{\left(n_{1j} + n_{2j}\right)^2 \left(n_{1j} + n_{2j} - 1\right)}$$

Basically, for any group (1 or 2), variance formulae are same. We can easily compute $\text{var}(O_i - E_i)$ as we have values for variables like d_{1j}, d_{2j}, n_{1j}, etc. Once we have everything, value of $(LogRank)_i$ can be compared with χ^2 table and corresponding p-value will decide acceptance or rejection of H_0.

Fortunately, we do not have to compute all these complex expressions as *lifelines* package provides a function for log-rank test. Function *logrank_test* takes two pairs of time and status values (one for each KM curves) and runs the hypothesis test like below:

```
from lifelines.statistics import logrank_test

results = logrank_test(veteran_df_1['time'], veteran_df_2['time'],
                       veteran_df_1['status'], veteran_df_2['status'])
results.print_summary()
```

t_0	-1
null_distribution	chi squared
degrees_of_freedom	1
test_name	logrank_test

	test_statistic	p	-log2(p)
0	0.01	0.93	0.11

Listing 4.12 **Perform log-rank test and get the result.**

We ran it for treatment groups 1 and 2 and got a high *p*-value (0.93 > 0.05). So, our null hypothesis is true, i.e., there is no significant difference between KM curves for treatment groups 1 and 2.

Analysis of Log-Rank Test

Some useful functional insights can be drawn from the results of log-rank test. Suppose we are running this test for one control and one treatment group to check the effectiveness of one drug. As discussed in Chapter 1, control group is a dummy group where a fake drug is given to simulate same mental effect on patients like an original drug, whereas treatment group gets the actual drug to be tested. Here, the objective is to check the effectiveness of the original drug on patients as compared to no drug given situation. Getting similar survival curves for two groups indicates that the drug is ineffective and is not able to improve survival probabilities of patients. However, there may be other detailed chemical and medical tests to declare a drug as ineffective and those are out of scope from our discussion. But, at least, what we can say is that log-rank test could be one of the important steps in a series for drug testing. Apart from having a control group, it can be tested for two treatment groups like we did for *veteran* dataset. In that case, two drugs are tested for their effectiveness. Acceptance of H_0 says that drugs are similar, and any one can be chosen.

Chapter 5

Models with Covariates

In previous chapters, our discussion revolved around a primary assumption about the survival distribution that it is only dependent on time. Certainly, it might have helped us to understand things easily, but practical scenarios are much more complex. It is observed in many cases that despite choosing a right probability distribution, desired accuracy is not obtained. Models are unable to capture some hidden variations in survival. It is caused by the external factors apart from time. Models discussed so far do not include these factors. Let us take the example of *veteran* dataset discussed in Chapter 2. There are factors like *celltype*, *karno*, *diagtime*, etc. which are not included while building the model. It may have (in fact, it has) some influence on the survival time. The dataset was also filtered with *trt=1*, i.e., treatment 1 condition. If we compare two survival time distributions for treatment groups 1 and 2 (basically one with treatment and another is control), some differences will be observed. These differences occur mostly due to the presence of explanatory variables such as *celltype* and others. These are also known as *covariates*. In this chapter, we will discuss two approaches of modeling survival distributions with explanatory factors, one directly with time and another with hazard.

Accelerated Life Model

For *veteran* dataset, survival times are differing for treatment groups 1 and 2. It can be observed that patients of one of the groups live longer (treatment group is supposed to live longer than control/test group). We can say that life span for one group is *accelerated* in this case. If the time variables are denoted by T_1 and T_0 for two groups, then

$$T_1 = \varphi T_0$$

where φ is a factor determining the acceleration. It can be said that time runs φ times faster for group 1 than group 0. One very practical example could be

DOI: 10.1201/9781003255499-5

comparing the survival time of a dog with that of a human. A dog's life span is much shorter than that of a human and time goes faster there for dog. So, the relationship could be $T_{\text{Dog}} = 7T_{\text{Human}}$, which says a human lives seven times more than a dog. A time interval of Δt in a dog's life span is equivalent to $7\Delta t$ in a human's lifespan. So, survival functions also will be affected due to this time scale change.

Now, if S_1 and S_0 are survival functions of two groups, then

$$S_1(t) = S_0(\varphi t)$$

Hence, for density function

$$\begin{aligned} f_1(t) &= -\frac{d}{dt} S_1(t) \\ &= -\frac{d}{dt} S_0(\varphi t) \\ &= -\varphi \frac{d}{d(\varphi t)} S_0(\varphi t) \\ &= \varphi \, f_0(\varphi t) \end{aligned}$$

Similarly, for hazard

$$h_1(t) = \varphi \, h_0(\varphi t)$$

This model is known as accelerated life model or accelerated failure time (AFT) model.

There may not be always two or more groups in the dataset. If there is single group's data, then can't we have S_0, h_0 and f_0? Of course, we can. These are known as *baseline survival, hazard* and *density* functions. It indicates initial survival state of the subjects without the influence of any covariates.

Now, the big question comes that how do we know the ideal value of φ? It is the effect of the covariates that determine φ. Weighted sum of covariates like a linear regression model should be a choice here. But there is a small problem. It should be positive as it works as multiplication factor on time (which itself is positive). And weighted sum may not be positive always as individual covariate x or the weight β could be negative. A better choice is to use $e^{\beta.x}$ as φ. It relieves us from the restrictions on x or β to be positive as $e^{\beta.x}$ is always positive irrespective of sign of $\beta.\boldsymbol{x}$. So, by mathematical notation,

$$\varphi = e^{\beta.x}$$

And,

$$\beta.x = \beta_1 x_1 + \beta_2 x_2 + . + \beta_k x_k$$
$$= \sum_{i=1}^{k} \beta_i x_i$$

There are k number of features in the survival dataset. $\beta.x$ is often denoted as η as is known as *risk score* of the model.

Now, the expressions of survival and hazard function become

$$S_1(t, \beta) = S_0\left(te^{\beta.x}\right)$$

$$h_1(t, \beta) = e^{\beta.x}\ h_0\left(te^{\beta.x}\right)$$

Survival and hazard are also functions of β now. As covariate x represents one subject (or patient in case of *veteran* dataset), then function $S_0(te^{\beta.x})$ basically represents survival function for that subject. Let us take an example of x_1 = {'squamous', 80, 9, 52, 10} from the dataset. There is a vector β having weights for each covariate. Let us assume hypothetically that *risk score* $\beta.x$ is evaluated as 2.5 (method of estimating β vector will be discussed later). So, the function for x_1 becomes like

$$S_{x1}(t) = S_0\left(te^{2.5}\right) = S_0(12.18t)$$

This looks interesting. It represents one subject x_1, but still having variable time t. If it is for a particular subject, then why is t coming? It is a survival probability distribution of over time for that subject x_1, where time t is the random variable. The dataset contains x_1 and its censoring status as 0 at time 25. Dataset does not contain any other record of that person at different time instance. So, there is no information about what happens next to that person. Survival function $S_{x1}(t)$ gives us that direction. As it is a probability distribution over time, we can compute what is the probability of survival of that person at time instance 35 (evaluated as $S_{x1}(35)$). Similarly, we can have distributions $S_{x2}(t)$, $S_{x3}(t)$, $S_{x4}(t)$ and so on for other persons.

In this context, let us try to understand significance of baseline survival function $S_0(t)$ mathematically as well as logically. We can see that

$$S_k(t) = S_0(t) = S_0\left(te^0\right)$$

where k is another record of a person. Basically, survival function of a subject (or person) whose *risk score* is zero is the *baseline survival* distribution (or function). This subject (in our case, k) may not exist. In our case, person k may not exist in *veteran* dataset whose *risk score* is zero. It is an imaginary or hypothetical record which makes the foundations of survival distributions of others.

For parametric models, S_1 and S_0, h_1 and h_0 must be from the same distribution family. If for any distribution, the relationship $S_1(t) = S_0(\varphi t)$ cannot be established, then we have to conclude that modeling with AFT is not possible for that distribution family. One thing to be noted here, for condition $S_1(t) = S_0(\varphi t)$ to be satisfied, parameter set may be different. We will now test this with *Weibull* distribution.

Weibull-AFT Model

Survival function for Weibull distribution is given by

$$S_0(t) = e^{-(\rho t)^{\kappa}}$$

Now,

$$\begin{aligned} S_0(\varphi t) &= e^{-(\rho\varphi t)^{\kappa}} \\ &= e^{-(\alpha t)^{\kappa}} \quad (\text{considering } \alpha = \rho\varphi) \end{aligned}$$

So, $S_0(\varphi t)$ follows the same pattern as $S_0(t)$, but with parameters α and κ. It is proved that *Weibull* distribution follows AFT model.

We can take one random survival distribution as $S_0(t) = t^{-k}$ and test its eligibility to be an AFT model.

Now,

$$\begin{aligned} S_0(\varphi t) &= (\varphi t)^{-k} \\ &= \varphi^{-k} t^{-k} \end{aligned}$$

Definitely, $S_0(\varphi t)$ is not in the same form as $S_0(t)$. So, in this case, $S_0(t) = t^{-k}$ does not follow the AFT model.

Determining Parameters β, κ, ρ for Weibull-AFT

We must take the same log-likelihood approach as we discussed in Chapter 3 for determining all parameters. Along with distribution parameter θ, we must estimate now covariate weight β also. So, following Equation (3.3), the composite generic log-likelihood for *AFT* model is given by

$$\begin{aligned} \log[L(\theta, \beta)] &= \sum c \log h(t, \theta, \beta) + \sum \log S(t, \theta, \beta) \\ &= \sum c\beta.\boldsymbol{x} + \sum c \log h_0\left(te^{\beta.x}, \theta\right) + \sum \log S_0\left(te^{\beta.x}, \theta\right) \end{aligned}$$

If we follow the above equation, then log-likelihood for Weibull-AFT model is given by

$$\log\left[L(\kappa,\rho,\beta)\right] = \sum c\beta.x + \sum c\log[\kappa\rho(\rho t e^{\beta.x})^{\kappa-1}] + \sum \log[e^{-\left(\rho t e^{\beta.x}\right)^{\kappa}}]$$
$$= \sum c\log\kappa + \sum c\kappa\log\rho + (\kappa-1)\sum\log(te^{\beta.x}) - \rho^{\kappa}\sum(te^{\beta.x})^{\kappa}$$
$$= \sum c\log\kappa + \sum c\kappa\log\rho + (\kappa-1)\sum(\log t + \beta.x) - \rho^{\kappa}\sum(te^{\beta.x})^{\kappa} \quad (5.1)$$

Technically, optimal values are given by

$$\left\{\widehat{\kappa},\ \hat{\rho},\ \widehat{\beta}\right\} = ar\,gmax_{\kappa,\rho,\beta}\left\{\log\left[L(\kappa,\rho,\beta)\right]\right\}$$

Now, we must solve three equations for finding tuple $\widehat{\kappa},\ \hat{\rho},\ \widehat{\beta}$.

$$\frac{\partial L}{\partial \kappa} = 0,\ \frac{\partial L}{\partial \rho} = 0,\ \frac{\partial L}{\partial \beta} = 0$$

Analytical solutions for these equations are too complex in form and it is difficult to get some concrete optimal values for parameters κ,ρ and β. Numerical solution (as discussed also in Chapter 3) is the only option we are left with.

We will fit this Weibull-AFT model on *veteran* dataset as discussed in Chapter 2. Let us check it again here.

	Unnamed: 0	trt	celltype	time	status	karno	diagtime	age	prior
0	1	1	squamous	72	1	60	7	69	0
1	2	1	squamous	411	1	70	5	64	10
2	3	1	squamous	228	1	60	3	38	0
3	4	1	squamous	126	1	60	9	63	10
4	5	1	squamous	118	1	70	11	65	10

Figure 5.1 Few records of the veteran dataset.

celltype is a categorical covariate. So, it must be converted to numerical values using one-hot-encoding scheme as shown here.

```
veteran_df_1 = veteran_df[veteran_df.trt==1].drop(['trt','Unnamed: 0'], axis=1)
veteran_df_1 = pd.get_dummies(veteran_df_1, columns=['celltype'])
veteran_df_1.head()
```

Listing 5.1 Apply one-hot encoder on the dataset.

And it produces the following output:

	time	status	karno	diagtime	age	prior	celltype_adeno	celltype_large	celltype_smallcell	celltype_squamous
0	72	1	60	7	69	0	0	0	0	1
1	411	1	70	5	64	10	0	0	0	1
2	228	1	60	3	38	0	0	0	0	1
3	126	1	60	9	63	10	0	0	0	1
4	118	1	70	11	65	10	0	0	0	1

Figure 5.2 **Output of one-hot encoder transformation.**

celltype gets decomposed into separate columns like *celltype_adeno*, *celltype_large*, etc. and each one is binary in nature. As we are interested with treatment group 1 only, dataset is filtered with *trt==1* condition. We will now do the actual work, i.e., fitting the Weibull model on this dataset.

```
from lifelines import WeibullAFTFitter

waft = WeibullAFTFitter(fit_intercept=False).fit(veteran_df_1, duration_col='time', event_col='status')
waft.print_summary()
```

Listing 5.2 **Fit Weibull model on the dataset.**

It produces the summary report of the model.

model	lifelines.WeibullAFTFitter
duration col	'time'
event col	'status'
number of observations	69
number of events observed	64
log-likelihood	-362.85
time fit was run	2020-11-29 07:03:26 UTC

		coef	exp(coef)	se(coef)	coef lower 95%	coef upper 95%	exp(coef) lower 95%	exp(coef) upper 95%	z	p	-log2(p)
lambda_	age	0.00	1.00	0.01	-0.02	0.03	0.98	1.03	0.19	0.85	0.24
	celltype_adeno	2.41	11.16	0.99	0.47	4.36	1.60	77.88	2.43	0.01	6.06
	celltype_large	3.64	38.01	0.96	1.76	5.52	5.81	248.81	3.80	<0.005	12.73
	celltype_smallcell	3.17	23.70	0.94	1.32	5.01	3.76	149.29	3.37	<0.005	10.38
	celltype_squamous	3.57	35.51	0.94	1.73	5.41	5.65	223.15	3.81	<0.005	12.79
	diagtime	-0.00	1.00	0.02	-0.04	0.04	0.96	1.04	-0.03	0.98	0.03
	karno	0.02	1.03	0.01	0.01	0.04	1.01	1.04	3.05	<0.005	8.77
	prior	-0.04	0.96	0.03	-0.10	0.02	0.91	1.02	-1.40	0.16	2.64
rho_	Intercept	0.08	1.08	0.09	-0.11	0.27	0.90	1.30	0.85	0.40	1.33

Concordance	0.69
AIC	743.69
log-likelihood ratio test	19.43 on 7 df
-log2(p) of ll-ratio test	7.17

Figure 5.3 **Output of the fitted Weibull model showing various parameters.**

It shows the covariates, their weights (*coef*) and corresponding p-values. A low p-value indicates that the weight of the covariate is less likely to be closer to zero, hence it has some contribution to the *risk score*. From this, it can be observed that *karno* is the most significant covariate as it has the lowest p-value. We will discuss this in detail in later sections of this chapter.

Now, we will do some more interesting discussions about the individual survival functions. In earlier sections, we discussed about the concept of different survival functions $S_{x2}(t)$, $S_{x3}(t)$, $S_{x4}(t)$, etc. for each data instance x_i. $S_{x_i}(t)$ is a probability distribution of survival for person x_i. We already built the model with Weibull distribution. Now, we can predict the survival probabilities of the same dataset with the built model and check the accuracy. Let's use function *predict_survival_function* and analyze its output.

```
estimated_prob_weibull = waft.predict_survival_function(veteran_df_1, times=veteran_df_1['time'])
estimated_prob_weibull.head()
```

Listing 5.3 Predict survival function using fitted Weibull model.

And it produces the following output:

	0	1	2	3	4	5	6	7	8	9	...	59	60	61	62
72.0	0.698056	0.648711	0.678640	0.565821	0.648564	0.329006	0.381687	0.808691	0.603313	0.760653	...	0.407520	0.813526	0.721699	0.774048
411.0	0.093206	0.057446	0.077370	0.023298	0.057360	0.000650	0.001732	0.246166	0.035584	0.164307	...	0.002669	0.256047	0.116129	0.184375
228.0	0.285586	0.221174	0.258837	0.137321	0.220999	0.020738	0.034806	0.476971	0.171744	0.385272	...	0.043734	0.486988	0.320753	0.409448
126.0	0.517305	0.452236	0.491228	0.351962	0.452048	0.130230	0.170998	0.677492	0.395903	0.605531	...	0.192816	0.684938	0.549886	0.625227
118.0	0.541234	0.477539	0.515778	0.378101	0.477354	0.149778	0.193024	0.695833	0.421887	0.626733	...	0.215867	0.702953	0.572916	0.645699

Figure 5.4 Survival function of the entire dataset (shown as probability distribution).

It looks interesting. *predict_survival_function* takes two inputs; one is the covariate set and other one is the series of time values for which we need predictions. We pass the time values from the original dataset itself. Column names 0, 1, 2, etc. in the output data frame are record numbers and row headers 72.0, 411.0, etc. are nothing but values of *time* column from the original dataset. For our convenience, we will transpose this and then try to understand its significance.

```
estimated_prob_weibull_T = estimated_prob_weibull.T
estimated_prob_weibull_T = estimated_prob_weibull_T.reset_index(drop=True)
estimated_prob_weibull_T.head()
```

Listing 5.4 Transpose probability distribution matrix.

And it produces the following output:

	72.0	411.0	228.0	126.0	118.0	10.0	82.0	110.0	314.0	100.0	...	12.0	260.0	200.0	156.0
0	0.698056	0.093206	0.285586	0.517305	0.541234	0.958542	0.661101	0.566125	0.169885	0.598624	...	0.949719	0.235781	0.337107	0.435730
1	0.648711	0.057446	0.221174	0.452236	0.477539	0.950300	0.607591	0.504102	0.118343	0.539144	...	0.939779	0.175603	0.270057	0.367820
2	0.678640	0.077370	0.258837	0.491228	0.515778	0.955363	0.639976	0.541406	0.147823	0.574999	...	0.945882	0.210508	0.309535	0.408231
3	0.565821	0.023298	0.137321	0.351962	0.378101	0.935120	0.519105	0.406018	0.060305	0.443559	...	0.921520	0.101363	0.178589	0.268177
4	0.648564	0.057360	0.220999	0.452048	0.477354	0.950275	0.607433	0.503922	0.118211	0.538970	...	0.939749	0.175443	0.269872	0.367628

Figure 5.5 **Transposed survival function matrix.**

Each column now becomes the random variable time t and each row is the probability distribution of survival for respective patients. If we observe the original record set, we can see that person 2 has status 1 at time instance 228. It may happen that there is no other record of this person at any other time instance in the dataset. But from survival probability distribution, we can get it. For example, we can see that at time instance 314.0, probability of survival for person 2 is 0.147821. We already discussed the mathematics behind it in earlier sections. We can now test the accuracy of estimated probabilities by *Brier Score* metric. We must tweak the Brier Score function as defined in Chapter 2 to incorporate the nature of the estimated probability output here.

```
import math
import numpy as np

def brier_score(actual, estimated):
    n = len(actual)
    total_error = 0.0
    for i in range(n): # iterate for each subject
        actual_prob = actual['status'].iloc[i]

        # take the computed survival probability for the given time
        estimated_prob = estimated.iloc[i][actual.iloc[i]['time']]

        if isinstance(estimated_prob, np.float64):
            error = (1.0 - estimated_prob) - actual_prob
        else:
            error = (1.0 - estimated_prob.iloc[0]) - actual_prob
        total_error = total_error + (error * error)

    return total_error/n
```

Listing 5.5 **Brier Score function (tweaked version).**

And the score is given by

```
brier_score(veteran_df_1, estimated_prob_weibull_T)
0.3506369133933204
```

Listing 5.6 **Get Brier Score using the function.**

Some explanation is needed for the tweaked version. Observe that, unlike univariate model, lot of subjects with different risk scores are involved here. For each subject, we have to pickup the actual censoring status & corresponding survival probability for a given time, then take square of the difference to sum up altogether. Remember that we must take the difference between two probabilities for the same time instance and it will come from the existing dataset.

Plotting Baseline vs Original Survival Function

We will now compute the baseline survival function $S_0(t)$. It has been discussed in earlier sections that for baseline survival function, *risk score*, i.e., βx is zero. We can see from the summary report that coefficients or weights are non-zero for almost all covariates. So, to make βx as zero, we can create a hypothetical record with all x_i as zero. Then it can be used to generate survival probability distribution with existing time values from the dataset. Function for that looks like below:

```
import numpy as np

def compute_baseline_survival_function(df, model, times=None):
    covariates = list(df[df.columns.difference(['time', 'status'])])

    # create a record with all covariate value x as zero
    X = pd.DataFrame(np.zeros(shape=(1,len(covariates))), columns=covariates)
    return model.predict_survival_function(X, times=times)
```

Listing 5.7 **Function to compute** *Baseline Survival function.*

Let us pick up now any random record, say record number 10, from the dataset.

```
veteran_df_1.iloc[10:11]
```

	time	status	karno	diagtime	age	prior	celltype_adeno	celltype_large	celltype_smallcell	celltype_squamous
10	42	1	60	4	81	0	0	0	0	1

Listing 5.8 **Pickup a sample record from the dataset (record number 10).**

We will plot both baseline and original survival functions for this record like below:

```
import matplotlib.pyplot  as plt

plt.plot(compute_baseline_survival_function(veteran_df_1, waft))
plt.plot(waft.predict_survival_function(veteran_df_1.iloc[10:11]))
plt.legend(["Baseline Survival", "Survival for X"])
```

Listing 5.9 **Plot baseline and original survival function in single graph.**

It produces the following output:

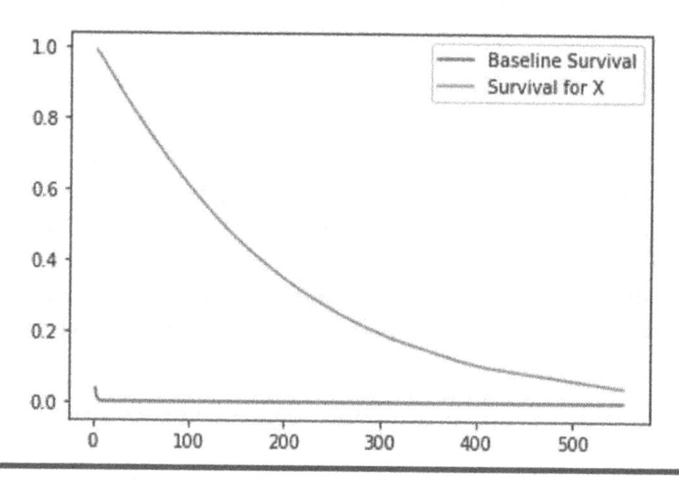

Figure 5.6 **Curves of baseline and original survival function.**

Baseline survival function is almost constant except some peak values at around 0th time. It may look odd, but it is true because baseline indicates a hypothetical record which may not exist.

Stepwise Computation of Relation $S_1(t, \beta) = S_0(te^{\beta.x})$

Now, we will see how the relation $S_1(t, \beta) = S_0\left(te^{\beta.x}\right)$ holds true. For this, we will consider another record at number 59.

```
veteran_df_1.iloc[59:60]
```

	time	status	karno	diagtime	age	prior	celltype_adeno	celltype_large	celltype_smallcell	celltype_squamous
59	12	1	40	12	68	10	0	1	0	0

Listing 5.10 **Pickup a sample record from the dataset (record number 59).**

We need a function to compute the *risk score* given the weights and covariate values.

```
def compute_risk_score(covariates, w):
    risk_score = 0.0
    for e in list(covariates):
        risk_score = risk_score + (w[('lambda_',e)] * list(covariates[e].values())[0])

    return risk_score
```

Listing 5.11 Function to compute risk score given X (covariate) values.

Reason for expression *w[('lambda_',e)]* will be clear if we see how we get weights from the summary.

```
weights = waft.params_.to_dict()
weights

{('lambda_', 'age'): 0.0023092822740220864,
 ('lambda_', 'celltype_adeno'): 2.412369978608446,
 ('lambda_', 'celltype_large'): 3.637943481764519,
 ('lambda_', 'celltype_smallcell'): 3.165353996646136,
 ('lambda_', 'celltype_squamous'): 3.569711534566083,
 ('lambda_', 'diagtime'): -0.0004652078957940859,
 ('lambda_', 'karno'): 0.024920405805181334,
 ('lambda_', 'prior'): -0.04099003976289921,
 ('rho_', 'Intercept'): 0.08015264014958057}
```

Listing 5.12 Show weight parameters of the survival model.

Now, it is time to compute the baseline survival function at one sample time instance $t = 333.0$.

```
import math

x = veteran_df_1[veteran_df_1.columns.difference(['time', 'status'])].iloc[59:60].to_dict()

t = 333.0
si = math.exp(compute_risk_score(x, weights))
compute_baseline_survival_function(veteran_df_1, model=waft, times=[t*si])
```

Listing 5.13 Compute baseline survival probability at a fixed time instance.

We compute $S_0\left(333.0e^{\beta.x}\right)$ above. It is the shifted baseline survival probability, and it is almost zero, i.e., the person almost has zero chance of survival at shifted time instance 26488. Now, we will compute survival function for record 59, i.e., $S_{x59}(333.0, \beta)$.

```
waft.predict_survival_function(veteran_df_1.iloc[59:60], times=[t])
```

	59
333.0	0.008936

Listing 5.14 **Compute original survival probability at the same time instance.**

It is also almost zero. So, it is proved that

$$S_{x59}(333.0, \beta) \approx S_0\left(333.0e^{\beta.x}\right)$$

Proportional Hazard Model

We saw that *accelerated life* model makes assumptions about the time scale. *Proportional hazard* (*PH*) model does the same but with respect to hazard only. It says hazard is affected by the covariates and is given by

$$h_1(t) = \varphi\ h_0(t)$$

$h_0(t)$ is the baseline hazard and φ is a constant and as usual estimated by $e^{\beta.x}$. We saw derivation of hazard equation as $h_1(t) = \varphi\ h_0(\varphi t)$ in case of *AFT* model. There is a subtle difference between this and PH model hazard equation. For AFT model, time is multiplied by φ and passed as input to the *baseline hazard*, whereas time is unchanged and passed as it is to the *baseline hazard* in case of PH model. It is clear from here that PH model does not affect time scale. For example, hazard rate or death rate of treatment 2 may be 3 times of treatment 1 for a subject (or patient in this case) in *veteran* dataset. It is explained by the PH model. Covariates associated with each subject influences the hazard rate.

Now, cumulative hazard for PH model is given by

$$\begin{aligned} H(t) &= \int_0^t h(u)du \\ &= \int_0^t \varphi h_0(u)du \\ &= \varphi H_0(t)(H_0(t) \text{ is baseline cumulative hazard}) \end{aligned}$$

So, baseline survival function is given by

$$S_0(t) = e^{-H_0(t)}$$

Then, survival function for PH model is given by

$$\begin{aligned} S(t) &= e^{-H(t)} \\ &= e^{-\varphi H_0(t)} \\ &= [e^{-H_0(t)}]^{\varphi} \\ &= [S_0(t)]^{\varphi} \end{aligned}$$

Like AFT model, definition of baseline survival is applicable here also, i.e., survival function of a person whose *risk score* is zero. We can see that

$$S_k(t) = \left[S_0(t)\right]^1 = \left[S_0(t)\right]^{e^0}$$

Hazard Ratio

Hazards of any two subjects can be expressed as ratio. For example, subjects i and j have covariate vectors $\boldsymbol{x}_i$ and $\boldsymbol{x}_j$ and, hazard functions h_i and h_j, respectively, then hazard ratio HR is given by

$$HR = \frac{h_i(t)}{h_j(t)} = \frac{h_0(t)e^{\beta x_i}}{h_0(t)e^{\beta x_j}} = e^{\beta(x_i - x_j)}$$

Above one can be more refined like below:

$$HR = e^{\beta_1(x_{i1}-x_{j1})+\beta_2(x_{i2}-x_{j2})+.+\beta_k(x_{ik}-x_{jk})}$$

Baseline hazard $h_0(t)$ gets cancelled out and HR is only dependent on covariates.

Hazard ratio is used to compare risks between two subjects. For any two subjects i and j, a value of HR > 1 indicates i has more hazard rate than j, and hence it is riskier and more prone to fall.

Cox-PH Model

Cox (1972) proposed this *PH* model with no assumption on $h_0(t)$. As we discussed earlier, parametric models make assumption about density functions and from there, survival and hazard functions are derived. In that sense, we can say assumption is there for hazard functions. Basically, Cox-PH is semi-parametric in nature. Baseline hazard and parameter estimations are done using a technique called *Breslow's* method.

Breslow's Method

As Cox-PH model does not assume any form of $h_0(t)$, we are left with estimating β only, i.e., weight vector of the covariates. This analysis starts with finding the conditional probability distribution of event sequence for subjects $i_1, i_2, \ldots, i_n$. If a subject i has an event at time t, then the conditional probability of the event given that at least one subject from the risk set also has an event at t can be obtained from Bayes rule and is given by

$$\frac{h_i(t)}{\sum_{k \in R} h_k(t)} = \frac{\varphi_i h_0(t)}{\sum_{k \in R} \varphi_k h_0(t)}$$

$$= \frac{\varphi_i}{\sum_{k \in R} \varphi_k}$$

R is the risk set which fails at t. $h_0(t)$ gets cancelled automatically and we can see that the probability is time-invariant. Like traditional Bayes rule, denominator term $\Sigma_{k \in R} h_k(t)$ is used as a normalizer factor to express the hazard as probability.

As we have got the conditional probability expression, total likelihood for all subjects is given by

$$L = \prod_{i=1}^{n} \frac{\varphi_i}{\sum_{k \in R} \varphi_k}$$

$$= \prod_{i=1}^{n} \frac{e^{\beta x_i}}{\sum_{k \in R} e^{\beta x_k}}$$

So, log-likelihood is given by

$$\log L = \sum \left[\beta x_i - \log \left(\sum_{k \in R} e^{\beta x_k} \right) \right]$$

And of course, the best estimate $\hat{\beta}$ is given by

$$\hat{\beta} = ar\,gmax_{\beta} \left\{ \log L \right\}$$

This $\hat{\beta}$ will be used to estimate baseline hazard.

A simple non-parametric estimate of total cumulative hazard $H(t)$ is the total number of events in the entire dataset and is given by

$$\hat{H}(t) = \sum_{t} c$$

where c is the censoring status (0 or 1) and it is working as counter here.

Now, $H(t)$ is summation of all $H_i(t)$s for each subject i and is given by

$$H(t) = \sum_{i=1}^{n} H_i(t)$$

We know $H_i(t)$ is the integral of $h_i(t)$; so, the same can be expressed as summation of a collection of $h_i(t)$s and is given by

$$\begin{aligned} H(t) &= \sum_{i=1}^{n} \sum_{t} h_i(t) \\ &= \sum_{i=1}^{n} \sum_{t} e^{\hat{\beta} x_i} h_0(t) \\ &= \sum_{i=1}^{n} e^{\hat{\beta} x_i} \sum_{t} h_0(t) \end{aligned}$$

Now, we can write

$$\sum_{i=1}^{n} e^{\hat{\beta} x_i} \sum_{t} h_0(t) = \sum_{t} c$$

And, of course, we get the expression of $h_0(t)$ as

$$h_0(t) = \sum_{t} \frac{c}{\sum_{n} e^{\hat{\beta} x_i}}$$

where $\hat{\beta}$ is the MLE of β.

$h_0(t)$ is independent of time t and does not assume any statistical prior distribution; hence, it is a non-parametric estimate. Once we get this $h_0(t)$, other functions like $h(t)$ and $S(t)$ can be easily computed by the standard formulae as discussed in earlier chapters. That is why, for any survival model, it is very important to estimate *baseline hazard* first.

We now model the same *veteran* dataset with Cox-PH approach.

```
from lifelines import CoxPHFitter

cxphft = CoxPHFitter(penalizer=0.01).fit(veteran_df_1, duration_col='time', event_col='status')
cxphft.print_summary()
```

Listing 5.15 Fit Cox-PH model using the veteran dataset.

And it produces weight details and p-values.

model	lifelines.CoxPHFitter
duration col	'time'
event col	'status'
penalizer	0.01
l1 ratio	0
baseline estimation	breslow
number of observations	69
number of events observed	64
partial log-likelihood	-200.27
time fit was run	2020-11-29 14:00:41 UTC

	coef	exp(coef)	se(coef)	coef lower 95%	coef upper 95%	exp(coef) lower 95%	exp(coef) upper 95%	z	p	-log2(p)
karno	-0.02	0.98	0.01	-0.04	-0.01	0.96	0.99	-2.76	0.01	7.45
diagtime	0.01	1.01	0.02	-0.03	0.04	0.97	1.05	0.26	0.80	0.33
age	-0.00	1.00	0.01	-0.03	0.02	0.97	1.02	-0.28	0.78	0.36
prior	0.04	1.04	0.03	-0.03	0.10	0.97	1.10	1.15	0.25	1.99
celltype_adeno	0.90	2.45	1.47	-1.98	3.77	0.14	43.44	0.61	0.54	0.88
celltype_large	-0.46	0.63	1.45	-3.30	2.39	0.04	10.88	-0.31	0.75	0.41
celltype_smallcell	0.12	1.12	1.44	-2.71	2.94	0.07	18.92	0.08	0.93	0.10
celltype_squamous	-0.31	0.73	1.45	-3.15	2.53	0.04	12.56	-0.21	0.83	0.27

Concordance	0.70
Partial AIC	416.55
log-likelihood ratio test	18.26 on 8 df
-log2(p) of ll-ratio test	5.69

Figure 5.7 Output of the fitted Cox-PH model showing various parameters.

From the details, we can see that *karno* is most important covariate as it has lowest p-value. Let us use the *predict_survival_function* and test the accuracy on the same dataset.

```
estimated_prob_coxph = cxphft.predict_survival_function(veteran_df_1, times=veteran_df_1['time'])
estimated_prob_coxph_T = estimated_prob_coxph.T
estimated_prob_coxph_T = estimated_prob_coxph_T.reset_index(drop=True)
brier_score(veteran_df_1, estimated_prob_coxph_T)

0.33438111698467665
```

Listing 5.16 Compute survival probabilities and accuracy (Brier Score).

Plotting Baseline vs Original Hazard Function

As it is a hazard-based assumption model, we will be interested more into baseline hazard functions and the relationship with original hazard. We know that $h_1(t) = \varphi\ h_0(t)$. So, if baseline hazard is multiplied with exponential of *risk score*, then our job is done. *lifelines* library does not provide any direct function like *predict_hazard* for Cox-PH model, but we can develop one using the above concept.

First, let us see the parameter structure of Cox-PH model.

```
weights = cxphft.params_.to_dict()
weights
```

```
{'karno': -0.02474350698811507,
 'diagtime': 0.005122246041946032,
 'age': -0.003643415245163211,
 'prior': 0.03611467419530616,
 'celltype_adeno': 0.8950691127961612,
 'celltype_large': -0.45549667183183695,
 'celltype_smallcell': 0.1175213146599877,
 'celltype_squamous': -0.3109690800967144}
```

Listing 5.17 Show weight parameters of the Cox-PH survival model.

It is much simpler than Weibull-AFT model. We should also tweak the *compute_risk_score* function accordingly.

```
def compute_risk_score(covariates, w):
    risk_score = 0.0
    for e in list(covariates):
        risk_score = risk_score + (w[e] * list(covariates[e].values())[0])

    return risk_score
```

Listing 5.18 Function to compute *risk score* for Cox-PH model.

Now, let us define *compute_hazard_function*.

```
import math

def compute_hazard_function(model, x, weights):
    si = math.exp(compute_risk_score(x, weights))
    return si * model.baseline_hazard_
```

Listing 5.19 Function to compute *hazard* for Cox-PH model for given covariates.

We can see that *compute_hazard_function* simply multiplies baseline hazard with exponential of the risk score as expected. *CoxPHFitter* class provides an output parameter known as *base_line_hazard*, which has all computed baseline hazard values from the time set provided in the original dataset.

Parameter x in the *compute_hazard_function* is a dictionary converted from a record like below:

```
x = veteran_df_1[veteran_df_1.columns.difference(['time', 'status'])].iloc[59:60].to_dict()
```

Listing 5.20 A sample record converted to a dictionary from the veteran dataset.

Record 59 is shown below:

```
veteran_df_1.iloc[59:60]
```

	time	status	karno	diagtime	age	prior	celltype_adeno	celltype_large	celltype_smallcell	celltype_squamous
59	12	1	40	12	68	10	0	1	0	0

Listing 5.21 A sample record (record number 59).

We can now plot the baseline vs original hazard function for record 59.

```
%matplotlib inline
import matplotlib.pyplot  as plt   plt

plt.plot(cxphft.baseline_hazard_)
plt.plot(compute_hazard_function(cxphft, x, weights))
plt.legend(["Baseline Hazard", "Hazard for X"])
```

Listing 5.22 Plot baseline and original hazard function in single graph.

And the output is

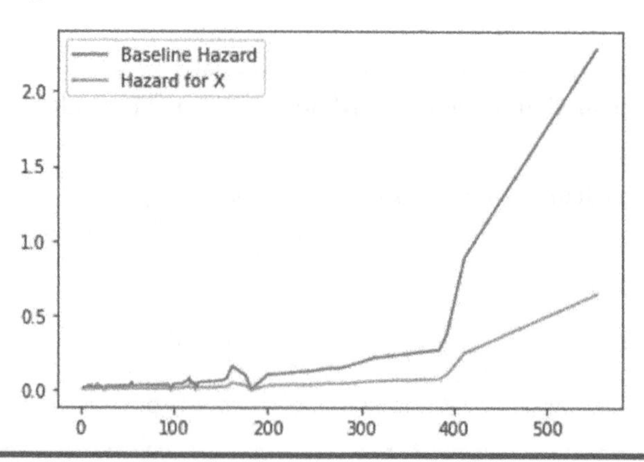

Figure 5.8 Curves of baseline and original hazard function.

It shows that hazard looks opposite to the survival function (it is discussed in Chapter 2 already).

Computing Hazard Ratio

We will consider two records and analyze which is more prone to fall.

```
veteran_df_1.iloc[59:61]
```

	time	status	karno	diagtime	age	prior	celltype_adeno	celltype_large	celltype_smallcell	celltype_squamous
59	12	1	40	12	68	10	0	1	0	0
60	260	1	80	5	45	0	0	1	0	0

Listing 5.23 **Two sample records (59 and 60) from veteran dataset.**

We will define a function *compute_hazard_ratio* for this.

```
def compute_hazard_ratio(x1, x2, w):
    risk_score_x1 = compute_risk_score(x1, w)
    risk_score_x2 = compute_risk_score(x2, w)
    return math.exp(risk_score_x1)/math.exp(risk_score_x2)
```

Listing 5.24 **Function to compute hazard ratio of two given records.**

We already discussed about how to compute hazard ratio and its formulae in earlier section. In above function, that concept has been implemented.

Now, we will extract covariates like before, prepare x59 and x60 accordingly and then invoke *compute_hazard_ratio.*

```
x59 = veteran_df_1[veteran_df_1.columns.difference(['time', 'status'])].iloc[59:60].to_dict()
x60 = veteran_df_1[veteran_df_1.columns.difference(['time', 'status'])].iloc[60:61].to_dict()

compute_hazard_ratio(x59,x60,weights)

3.680115704521152
```

Listing 5.25 **Computed hazard ratio between records 59 and 60.**

We can conclude that person 59 is more prone to fall as compared to person 60.

Weibull–Cox Model

It is the model where hazard is derived from *Weibull* density function. We already discussed about Weibull distribution for AFT model. For Weibull–Cox, assumption starts with the hazard like below:

$$\begin{aligned} h(t) &= e^{\beta.x} h_0(t) \\ &= e^{\beta.x} \kappa\rho(\rho t)^{\kappa-1} \\ &= e^{\beta.x} \rho\kappa t^{\kappa-1} \\ &= \lambda\kappa t^{\kappa-1} \; (\textit{taking } \lambda = e^{\beta.x}\rho) \\ &= \kappa\lambda(\lambda t)^{\kappa-1} \end{aligned}$$

So, we can see that $h(t)$ is following a Weibull hazard function with parameter κ and λ. Basically, λ is a result of re-parameterization.

Weibull (and *exponential* also) is the only distribution that follows both AFT and PH model. No other distribution has this property.

Determining Parameters β, κ, ρ for Weibull–Cox

Maximum likelihood estimates (MLE) of β, κ and ρ can be found in a similar manner like Weibull-AFT model. Likelihood expression is given by

$$\begin{aligned} \log\,[L(\beta,\kappa,\rho)] &= \sum c \log h(t,\beta,\kappa,\rho) + \sum \log S(t,\beta,\kappa,\rho) \\ &= \sum c \log[e^{\beta.x} h_0(t,\kappa,\rho)] + \sum \log[S_0(t,\kappa,\rho)]^{e^{\beta.x}} \\ &= \sum c\beta.\boldsymbol{x} + \sum c \log h_0(t,\kappa,\rho) + \sum e^{\beta.\boldsymbol{x}} \log S_0(t,\kappa,\rho) \end{aligned}$$

Putting actual Weibull distribution expressions in the above equation and rearranging, we get

$$\begin{aligned} &\log\left[L(\beta,\kappa,\rho)\right] \\ &= \sum c\beta.\boldsymbol{x} + \sum c \log\kappa + \sum c\kappa\log\rho + (\kappa-1)\sum \log t - e^{\beta.\boldsymbol{x}}\rho^{\kappa}\sum t^{\kappa} \end{aligned}$$

Like earlier, to find $\hat{\beta}$, $\widehat{\kappa}$ and $\hat{\rho}$, we need to solve three partial derivative equations, respectively,

$$\frac{\partial \log L}{\partial \beta} = 0, \; \frac{\partial \log L}{\partial \kappa} = 0, \; \frac{\partial \log L}{\partial \rho} = 0$$

As usual, closed form analytical solution of the above three equations is difficult; hence numerical solution for β, κ and ρ is recommended.

Significance of Covariates

So far, we have included all covariates while building the model. A major question is : which of them is most important? Or better to say, which covariate influences hazard or survival probability maximum? This can be answered by running a few statistical significance tests as discussed next.

Wald Test

It is the most commonly used test. It tests whether weight of any covariate is zero or not. A zero weight indicates that the covariate is not at all important and has no contribution in score function. But a complete zero weight may not be always there. So, a statistical test is necessary to check how much close it is to zero, i.e., how much statistically significant it is. It is done by Z-test and the statistic is given by

$$Z_w = \frac{\hat{\beta}}{SE\left(\hat{\beta}\right)}$$

$\hat{\beta}$ is the MLE of β. *SE* stands for *standard error* of weight $\hat{\beta}$. SE is nothing but the probable error while estimating β. We know that Z-test compares two sample statistics. In our case, we are comparing $\hat{\beta}$ with 0. It is actually

$$Z_w = \frac{\hat{\beta} - 0}{SE\left(\hat{\beta}\right)}$$

Here the null hypothesis is H_0: $\beta = 0$ against alternative H_1: $\beta \neq 0$. We can reject H_0 if $|Z_w| > Z_{\alpha/2,}$ where α is the confidence interval. Generally, it is kept as 95%. In common terms, p-value of Z_w should be more than 0.05 to be significant.

Now, let us take the same example of veteran dataset and use the fitted Cox-PH model there to see the weights of covariates.

model	lifelines.CoxPHFitter
duration col	'time'
event col	'status'
penalizer	0.01
l1 ratio	0
baseline estimation	breslow
number of observations	69
number of events observed	64
partial log-likelihood	-200.27
time fit was run	2020-11-29 13:51:53 UTC

	coef	exp(coef)	se(coef)	coef lower 95%	coef upper 95%	exp(coef) lower 95%	exp(coef) upper 95%	z	p	-log2(p)
karno	-0.02	0.98	0.01	-0.04	-0.01	0.96	0.99	-2.76	0.01	7.45
diagtime	0.01	1.01	0.02	-0.03	0.04	0.97	1.05	0.26	0.80	0.33
age	-0.00	1.00	0.01	-0.03	0.02	0.97	1.02	-0.28	0.78	0.36
prior	0.04	1.04	0.03	-0.03	0.10	0.97	1.10	1.15	0.25	1.99
celltype_adeno	0.90	2.45	1.47	-1.98	3.77	0.14	43.44	0.61	0.54	0.88
celltype_large	-0.46	0.63	1.45	-3.30	2.39	0.04	10.88	-0.31	0.75	0.41
celltype_smallcell	0.12	1.12	1.44	-2.71	2.94	0.07	18.92	0.08	0.93	0.10
celltype_squamous	-0.31	0.73	1.45	-3.15	2.53	0.04	12.56	-0.21	0.83	0.27

Figure 5.9 Summary of Cox-PH model.

lifelines library has already calculated the *p*-values of all covariate weights. We can see that *karno* has a low *p*-value (<0.005), and hence it is highly significant covariate.

Likelihood Ratio Test

It tests that if we set any covariate weight to zero then does that change the overall likelihood significantly or not? For example, if setting *karno* attribute (covariate) to zero (in *veteran* dataset) creates some significant effect on likelihood, then it is an important covariate. A change in overall likelihood ultimately causes some changes in survival probability. So, in that sense, a covariate can influence (increase or decrease) survival and hazard.

Test statistic of *likelihood ratio test* for a covariate β is given by

$$LR_\beta = 2\left[L\left(\beta = \hat{\beta}\right) - L(\beta = 0)\right]$$

where $\hat{\beta}$ is the *MLE* of β. From the expression, we can see that it is nothing but the difference between two log-likelihood values: log-likelihood by setting β to its best estimate and setting it to zero. A bigger value of LR_β indicates more significance for the covariate. But what value is sufficient? That will be determined by a statistical test. There would be a null hypothesis as β is zero versus the alternative as β is non-zero. As LR_β follows a χ^2 distribution with 1 degree of freedom, the null hypothesis can be accepted or rejected depending on the *p*-value. Generally, any *p*-value less than 0.05 results in rejection of the null hypothesis, i.e., it can be concluded that the covariate is important.

Selection of Covariates

Very often for modeling a survival analysis problem, we are interested to know the important covariates that should be chosen to build the model. In fact, some cases are built only to know the set of important covariates. Like lung cancer analysis use case (*veteran* dataset), health experts/doctors may be interested to know only those factors which can influence the disease most. It may help in discovering new medicines. So, a right procedure/algorithm is needed that can tell us which covariate we should choose to build the model. Another objective of doing this is to make a perfect model and get rid of not so important factors which make the model complex and increase the building time. We will now discuss one of the common techniques of choosing covariates.

This technique is known as forward selection procedure. It starts with univariate models, one for each covariate. The model with the smallest *p*-value from the

LR test is chosen. Then, with that covariate included, a series of separate models fitted with each single additional covariate are tested and the best one is chosen as usual. Reason for choosing smallest p-value can be explained. We saw the expression of LR_β for testing H_0: $\beta = 0$, but this can be extended to test whether a collection of weight vector $\boldsymbol{\beta} = 0$ or not. It says whether all of weights are close to zero or not. A smaller p-value indicates chances of rejecting H_0 which yields a better model. Iterations go on until the p-value reaches a critical limit. It is better explained by the following step-wise algorithm.

Forward Selection Algorithm

Step 1: Initialize global set S with all covariates

Step 2: For each x_i under S, add this to the existing model and build a series of updated models $\{M_i\}$

Step 3: Run *likelihood ratio test* against null model M_0 (where all covariate weights are assumed to be zero) with all models built at *Step 2*

Step 4: Pick up the model $M_{\min}$ which has minimum p-value

Step 5: Identify the extra covariate x_i which was added to $M_{\min}$ at *Step 2* and remove that from global set S

Step 6: Go to *Step 2* if S is not empty or p-value of $M_{\min}$ is still greater than some threshold (5% or 10%); otherwise declare $M_{\min}$ as final model.

We will implement this algorithm by two functions. First one is *build_optimal_model_forward_selection* which iteratively uses second utility function *_get_next_best_covariate_* to try out various combinations and get the optimal p-value and model. It also performs task mentioned in *Step 2* and looks like as given below:

```
def build_optimal_model_forward_selection(df, time_column, event_column, threshold_p_value):
    current_covariates = []
    global_set_covariates = list(df[df.columns.difference([time_column, event_column])])
    optimal_model = None

    while True:

        # Get next best covariate by running a combination of models
        best_covariate, min_p_value, current_optimal_model = _get_next_best_covariate_(X=df,
                                                    time_column=time_column,
                                                    event_column=event_column,
                                                    current_covariates=current_covariates,
                                                    remaining_covariates=global_set_covariates)

        # Add to the current set of covariates and remove from global set
        current_covariates.append(best_covariate)
        global_set_covariates.remove(best_covariate)
        optimal_model = current_optimal_model

        if min_p_value <= threshold_p_value or len(global_set_covariates) <= 0:
            break

    return optimal_model
```

Listing 5.26 Function to build an optimal model with forward selection.

Function *_get_next_best_covariate_* picks up one covariate at a time, joins to the current list and builds a series of models. It keeps the model which has minimum *p*-value. The function looks like below:

```
import sys

def _get_next_best_covariate_(X, time_column, event_column, current_covariates=None, remaining_covariates=None):
    min_p_value = sys.float_info.max
    best_covariate = None
    min_p_model = None

    # Iterate over the set and fit a model with additional covariate
    for covariate in remaining_covariates:
        covariates = []
        covariates.extend(current_covariates)
        covariates.append(covariate)
        covariates.append(time_column)
        covariates.append(event_column)

        model = CoxPHFitter(penalizer=0.01).fit(X[covariates], duration_col=time_column, event_col=event_column)

        current_p_value = model.log_likelihood_ratio_test().p_value

        # Find minimum of the p-values
        if min_p_value > current_p_value:
            best_covariate = covariate
            min_p_value = current_p_value
            min_p_model = model

    return best_covariate, min_p_value, min_p_model
```

Listing 5.27 Function to get the next best covariate (to be used internally).

And finally, both can be utilized like below to get the optimal model.

```
optimal_model = build_optimal_model_forward_selection(df=veteran_df_1,time_column='time',
                                                      event_column='status', threshold_p_value=0.00189)
optimal_model.print_summary()
```

Listing 5.28 Use *build_optimal_model_forward_selection* to build the model.

The output is

model	lifelines.CoxPHFitter
duration col	'time'
event col	'status'
penalizer	0.01
l1 ratio	0
baseline estimation	breslow
number of observations	69
number of events observed	64
partial log-likelihood	-202.53
time fit was run	2020-11-27 06:25:59 UTC

	coef	exp(coef)	se(coef)	coef lower 95%	coef upper 95%	exp(coef) lower 95%	exp(coef) upper 95%	z	p	-log2(p)
karno	-0.03	0.97	0.01	-0.04	-0.01	0.96	0.99	-3.38	<0.005	10.42
celltype_adeno	0.98	2.65	0.38	0.22	1.73	1.25	5.64	2.54	0.01	6.48

Concordance	0.66
Partial AIC	409.05
log-likelihood ratio test	13.75 on 2 df
-log2(p) of ll-ratio test	9.92

Figure 5.10 Summary of optimal Cox-PH model.

Explainability of Models

At high level, primary objective of a statistical model is to provide forecasting or prediction and these are mostly black box in nature. For an end user, it is very difficult to comprehend what's going on behind. For instance, a Cox-PH model gives us a list of covariate weights with all p-values. From the weights, we can draw some insights about the effect of covariates. A p-value just says whether a covariate is important or not but does not say anything about its concrete effect on the model. We need little more explainability here. It is applicable for both AFT & PH based models. Remember the expression of survival function of AFT model,

$$S(t) = S_0\left(te^{\beta.x}\right)$$

A positive value of risk score $\beta.x$ will produce a higher value for the term $e^{\beta.x}$ and positive weights of covariates will push the risk score towards positive side. As the survival function is monotonically decreasing over time, following relation holds true for it:

$$S_0(at) < S_0(bt)$$

where $a > b$. What it means that covariate with positive weight causes the survival probability to decrease for AFT based model (given that the value of covariate x is also positive). Just opposite is true for covariates with negative weight.

Consider now the expression of PH model,

$$S(t) = \left[S_0(t)\right]^{e^{\beta.x}}$$

A positive value of risk score $\beta.x$ will result multiple $S_0(t)$ s multiplied together which in turn will reduce the value of $S(t)$. So, by a similar logic like AFT model, positive weights contribute to risk score to decrease the survival probability and negative weights increase it.

Now, we will see the explainability in effect for the practical use case as shown in Figure 5.10. Observe that only two covariates *karno* and *celltype_adeno* are the important ones given that we maintain cut-off *p*-value at 0.00189. It, of course, depends on what level of significance we want. As *celltype_adeno* has positive weight, it contributes a positive component in *risk score* that causes its increment. So, it can be concluded that patients with cell type *adeno* have more hazard value and less chances of survival. Similar conclusions can be drawn for other covariates also if for once we ignore their p-value and consider all of them as important. It can produce very meaningful insights in analysis of effect of drugs, material-life testing, rare event discovery etc. For example, in drug testing, model explainability can show which factors (covariates) improve survival chances after the application of the drug on patients.

Index

Note: Page numbers in *italics* refer to figures

F

G

H

I

K

L

M

N

P